José Gustavo da Silva Melo
Elisabeth C. Silva
Paulo R. Sarmento

Study of environmental degradation caused by socio-spatial actions

José Gustavo da Silva Melo
Elisabeth C. Silva
Paulo R. Sarmento

Study of environmental degradation caused by socio-spatial actions

ScienciaScripts

Imprint
Any brand names and product names mentioned in this book are subject to trademark, brand or patent protection and are trademarks or registered trademarks of their respective holders. The use of brand names, product names, common names, trade names, product descriptions etc. even without a particular marking in this work is in no way to be construed to mean that such names may be regarded as unrestricted in respect of trademark and brand protection legislation and could thus be used by anyone.

Cover image: www.ingimage.com

This book is a translation from the original published under ISBN 978-613-9-65142-9.

Publisher:
Sciencia Scripts
is a trademark of
Dodo Books Indian Ocean Ltd. and OmniScriptum S.R.L publishing group

120 High Road, East Finchley, London, N2 9ED, United Kingdom
Str. Armeneasca 28/1, office 1, Chisinau MD-2012, Republic of Moldova, Europe
Printed at: see last page
ISBN: 978-620-7-85543-8

We dedicate this work to our families, friends and teachers who have contributed so much to our academic and personal success and growth.

ACKNOWLEDGEMENT

We thank God, because without him we wouldn't be here and in him we place our faith, our trust and our love. To our parents, for all their patience, unconditional love and for all the times we thought we'd throw everything away and they made us change our minds.

To our friends, who in difficult times helped us realise that companionship overcomes all the barriers in our paths. This requires trust. It's true that sometimes we don't understand each other very well, but we haven't forgotten the years of hardship we went through to reach the end of the dream.

Finally, we would like to thank everyone for the opportunity they gave us, when they helped to expand our modest knowledge of the world, as well as the satisfaction we got from travelling around the premises of the Federal Institute of Pernambuco during the years that passed so quickly, leaving us longing. So thank you very much and may God bless us. Amen.

SUMMARY

To analyse the socio-environmental degradation caused by the Paulo Afonso hydroelectric complex, through the socio-spatial consequences caused by the construction of the energy project. Hydroelectric power stations in Brazil have been a driving force behind the country's industrial development. This process began in the 1950s under the government of Juscelino Kubitscheck, who prioritised the expansion of Brazil's energy matrix based on water resources. The territorial area of Paulo Afonso belongs to homogeneous micro-region 147, Sertão de Paulo Afonso, occupying a territorial area of 1,545.192 km^2. The problem addressed in this research is relevant because of the need to highlight the complex nature of the factors that explain socio-spatial relations, in terms of the evolutionary conditions of the process of setting up hydroelectric plants in the São Francisco sub-middle. On the other hand, the dams boosted the country's development and offered new prospects for the north-eastern region of Brazil. Since then, Paulo Afonso/BA has found itself in a new reality as a result of the socio-spatial reorganisation caused by the Paulo Afonso hydroelectric complex. The research strategies, in terms of applicability, can be classified as: qualitative and quantitative approach, since it uses a very broad layout. It should be emphasised that this work has greater academic relevance because of the city's importance to the northeast, its strategic role in the process of industrial acceleration in this region and the preservation of the city's memory. The results show that the Paulo Afonso hydroelectric complex, made up of the Paulo Afonso I, II, III, IV and Apolônio Sales (Moxotó) power stations, produces 4 million, 279,000 and 600 kW. That the state comes to be **identified as Brazilian "society" in general, and the real societies are just** objects of the state. The project becomes a natural and inevitable event. Also noteworthy are the socio-environmental impacts produced by the construction of the hydroelectric dam, which required the relocation of around 64,000 people. Finally, it can be concluded that harnessing the hydroelectric potential of the São Francisco represented a process for northeastern society of introducing a model for inducing development that was widespread at the time. In other words, development reduced to the idea of simple economic growth.

Keywords: Social impacts; Regional development; Hydroelectric power; Historical heritage.

SUMMARY

CHAPTER 1

INTRODUCTION

The struggle between man and nature has always been a constant in the north-eastern region of Brazil. The force of nature's elements has always demanded a certain amount of effort from those who inhabit it. As far as we know, the first inhabitants of the Paulo Afonso region were Amerindian peoples, many of whom came from the coast fleeing the Portuguese, who "discovered" the São Francisco River on 4 October 1501. For the Indians, the great river was "Opará", which means "river-sea". The colonisers baptised it with the name "São Francisco", because it was discovered on the day of the Catholic saint. Today, however, there are few indigenous reserves in Paulo Afonso/BA. The Raso da Catarina ecological reserve is currently home to the Pankararé tribe (FOLHA SERTANELA, 2015, p. 6).

However, some records state that on 3 October 1725 the sertanista Paulo de Viveiros Afonso received a sesmaria in the lands of the province of Pernambuco, whose limits reached the waterfalls known as "Cachoeira Grande", "Forquilha" (due to its shape) or "Sumidouro". Before this date, there is no record in Brazil or Portugal that mentions the waterfall under the name of Paulo Afonso. The sesmeiro is said to have founded a small tapera on the Bahian side of the land known as "Tapera de Paulo Afonso", where the Centenário neighbourhood is today, which would have been the city's first residential area (MUCCINI; MALTA, 2007).

This region has always served as a route for travellers, crossing the river at the town of Santo Antônio das Glórias, now "Nova Glória", of which Paulo Afonso was a part (and which was only emancipated in 1958), where the "route of the oxen" passed during the colonisation period, "Curral dos bois" was the name given to the region and "rio dos currais" was how the São Francisco was called during this period, which served as a landing place for cattle on long journeys. Travellers and their cattle quenched their thirst and tiredness on the banks of the river, which was the path of the settlers who populated and cultivated the backlands (SANTIN; FLORES, 2006).

Nevertheless, the historical approach of the São Francisco Hydroelectric Company, in order to develop a socio-environmental analysis, through the socio-spatial contributions arising from the implementation of hydroelectric plants in the area surrounding the municipality of Paulo Afonso, Bahia, Brazil, is one of the ways to guide the study. We will therefore focus on the historical facts that marked the pioneering era, with an emphasis on the references researched through authors who tell the story, briefly summarising the ideas behind the pioneering use of the "Paulo Afonso waterfalls" and the creation of CHESF.

In this area, where the states of Bahia, Pernambuco, Alagoas and Sergipe are geographically located (as well as being very close to other northeastern states such as Ceara and Paraíba), the São Francisco River is of strategic importance, as it helps to reduce the aggravating poverty rates in what is one of the driest and poorest areas of Brazilian territory, and is located in the so-called "drought polygon" of the upper northeastern hinterland (MUCCINI; MALTA, 2007).

However, the region's biggest attraction has always been its geographical features, such as the canyons and the Paulo Afonso Waterfall. The poet Castro Alves, who never got to see it, dedicated a poem to it, "Cachoeira de Paulo Afonso", which was the title of one of his books. On 20 October 1859, the then Emperor Pedro II and his entourage also visited the much-talked-about Cachoeira (REIS, 2004).

However, as it was part of the Cangaço trail, the Cachoeira area also served as a hiding place for Lampião. The caves in the canyon, known as the "Fumas do Morcego" (Bat's Caves), sheltered the cangaceiro and his gang, a fact that, according to historian Antônio Galdino, is still disputed (CME, 1998; 1993).

Therefore, the story is not just a technical one, i.e. of a major technical undertaking, but mainly reveals what is most human about the social relations that were established in this region (PEREIRA, 2012; PESAVENTO, 2007).

Therefore, in order to achieve this goal of analysing social relations, we resorted to conceptual and methodological tools, both documental and technical, which are so valuable in view of the objectives of this work, which favoured content analysis of the various social agents in the pioneering history of Paulo Afonso and the São Francisco Hydroelectric Company (CHESF), reporting facts and cases, incorporating the culture brought by each of them (TEIXEIRA, 2005). We emphasise that these pioneers came from different cities, states and even countries, adding their cultures to those of other people who were here looking for work (MENDONÇA; BRITO, 2007).

So, despite the uncertain days in 2017, the current CHESF employees and those pioneers from 1948 to the end of 1954 and beginning of 1955, and then in the following decades, in a total of almost 50 years of uninterrupted work, have every right to rejoice in the beautiful history they helped to build in the region, digging tunnels, building dams, fighting with the impetuous river to generate and transmit the electricity that gradually changed the face of the entire Northeast region (FOLHA SERTANELA, 2015, p. 7).

In view of this, it's fair to say that there are two Northeastern countries: the one before and the one after Chesf. Sixty years ago, when the heroic founders received the President of the Republic, Getúlio Vargas, and the country's highest authorities to officially inaugurate the Paulo Afonso Power Station, Brazil was also going through moments of political crisis (RAMPAZO; ICHIKAWA, 2013).

However, in Brazil, electricity production gained importance with the industrialisation process in the 1950s and 1960s. During this period, with the implementation of the Target Plan[1] (1956-1961) by the Juscelino Kubitschek government, the demand for energy increased (SILVA, 2011).

With the construction of the Paulo Afonso I hydroelectric power station in 1954, Brazil began a phase of large-scale hydroelectric projects. This phase continued with the construction of the Furnas, Urubupungá, etc. plants throughout the 1960s (FURNAS, 2007). This phase was characterised by the so-called "developmentalist" model, in which hydroelectric plants were created to ease the demand for energy generated by the emerging industry and as a source of new jobs, without questioning the impacts caused (FOLHA SERTANELA, 2015, p. 7).

In the military period (1964-1984), continuing this model based on the military's developmentalist ideology, a large number of hydroelectric power stations were built, as Sevá (2008) states. In order to keep up with the capitalist development model, Brazil needed to increase its energy capacity. Hence the construction of numerous hydroelectric power stations during this period (ROSA, 2007).

---

[1] Plano de Metas (Target Plan): Development programme implemented by the Juscelino Kubitschek government (19561961), which involved major state investments in various sectors of the economy, agriculture, education, health, energy, transport, mining and construction. The government's aim was to make the country **grow "50 years in 5" (SILVA, 2011; CME, 1998).**

As a result, the state electricity sector's policy began to prioritise energy production through the construction of hydroelectric plants. This type of energy model implemented in the country, based on the construction of large hydroelectric plants, caused serious damage to the environment and the populations affected (GAMEIRO, 2012).

The population affected by the construction of the power plants was relocated and resettled elsewhere, giving rise to various resettlements, the pioneers being Nova Glória, Quixabá and Santa Brígida, in the state of Bahia. These resettlements were set up by CHESF (Companhia Hidroelétrica do São Francisco) to mitigate the impacts caused by the filling of the lake at the Paulo Afonso I, II, III, IV and Apolônio Sales HPPs (ANEEL, 2005).

As a result, the formation of hydroelectric dam reservoirs generally affected more fertile soils and arable land, disintegrating the local population, which lost its historical characteristics, cultural identity and relationship with the place, as well as altering aquatic ecosystems and destroying flora and fauna (SEVÁ, 2008).

In this way, the construction of hydroelectric plants caused irreversible socio-environmental impacts in various parts of the country (Bahia, Sergipe, Alagoas, São Paulo, Pará, etc.), including some areas of the state of Bahia, including the municipality of Paulo Afonso (LOUREIRO, 2012; 2003). In this part of the state, while on the one hand the power plants played an important role in regional dynamics, especially during the period of their construction, creating jobs and increasing the population in the municipalities involved, on the other hand, they caused social and environmental impacts, causing the riverside and island populations to leave, as they practised subsistence and commercial agriculture, fishing, among other activities (SOUZA, 2008).

At the same time, the state played a significant role in driving industrialisation, not only through its role as a provider of public goods, but also, and above all: a) in defining, coordinating and financially supporting large blocks of investments that determined the main changes in Brazil's economic structure; b) in building infrastructure, with the aim of integrating the road, energy, urban and telecommunications systems; c) in the direct production of intermediate inputs that were indispensable for heavy industrialisation CARLOS, 2011).

These territorial policies developed through development plans were aimed at integrating the Brazilian regions and correcting regional disparities, especially in the Northeast, as well as the economic occupation of the Amazon, through industrial hubs as a way of decentralising the industrial structure. This framework of investment in infrastructure also favoured the demand for energy (SILVA, 2011).

As a result of foreign indebtedness and high interest rates, construction work and the development of new hydroelectric projects were paralysed. As a result, Brazil's energy programme became unviable for decades (CANO, 2008).

However, the still rather timid resumption of investment in the sector in 1997/1998 became possible due to the renegotiation of foreign debts, the enactment of a set of resolutions and environmental laws and, above all, changes to the Brazilian constitution, which began to allow private capital to participate and invest in the generation and distribution of electricity. The way forward now is the progressive privatisation of the sector, given the state's inability to continue operating in the productive sector (SILVA, 2013).

In the 1990s, following the international neoliberalism, which advocated the weakening of the state, i.e. the minimal state with specific characteristics for underdeveloped countries. In Brazil, the wave to privatise basic infrastructure sectors (health, education, telecommunications, transport and energy) through various measures adopted, despite all the effort of previous decades dedicated to building this same infrastructure with public resources (ACSELRAD, 2007).

With neoliberalism, the state is now playing the role of mediator of social and economic policies and is no longer an entrepreneur, as it was in the 1950s to 1970s with the developmentalist model. Today, health, education, social security, energy, transport and communications are increasingly managed by the private sector (CANO, 2008).

According to data presented by Zhouri and Oliveira (2007), in Brazil more than one million people have been compulsorily displaced due to the flooding of their land by hydroelectric dams. This change is not just in physical space, but mainly in social relations, labour occupations, routines, symbolic representations, bonds, in other words, the identity of these populations. And therein lies the difficulty for the populations in adapting to this new situation and re-signifying their identities, linked to the past territorial space left for the construction of the new dam reservoir.

Finally, there is agreement among authors (HABERMAS, 2009; CASTELLS, 2008; ACSELRAD, 2007; ZHOURI; OLIVEIRA, 2007) regarding the conception of socio-spatial processes, where there is a personal, primary core that, when socialised with other individuals in society, undergoes new identifications and forms its identity. In this sense, the self-reflexive nature of the individual's actions and the space-time continuum stand out (HABERMAS, 2009; CASTELLS, 2008; ERIKSON, 1987; CASTELLS, 1983).

1.1 Delimitation of the Research Problem

The construction of hydroelectric power stations has caused irreversible socio-environmental impacts in various parts of the country, including some areas of the state of Bahia, including Delmiro Gouveia, now the municipality of Paulo Afonso. These power stations played an important role in regional dynamics, especially during the period of their construction, with the creation of jobs and an increase in population in the municipalities involved (HAMBUS, 2012). On the other hand, they had socio-environmental impacts, causing the riverside and island populations to leave, as they practised subsistence and commercial agriculture, fishing, among other activities (SOUZA, 2008; ZHOURI; OLIVEIRA, 2007).

In view of the above, the problem addressed in this research has its relevance linked to the need to highlight the complex nature of the factors that explain socio-spatial relations, in terms of the conditions of evolution of the process of implementing hydroelectric dams in the São Francisco sub-middle.

However, beyond this context, the influence of the municipality of Paulo Afonso/BA in the socio-spatial and environmental articulation of the area was highlighted, both on an interstate and inter-municipal scale in the process of "development" of the region under study.

1.2 Justification

We used the conceptual and methodological tools of Oral History, which is so valuable in view of the objectives of this work, which favoured the life story narratives of the various social agents in the pioneering history of Paulo Afonso, to relate facts and cases, incorporating the culture brought by each of them, as well as the documents, archives and memories of the São Francisco Hydroelectric Company (CHESF), in addition to academic documents and the vast bibliography on the municipality of Paulo Afonso/BA (SOUZA, 2011; ALBERTI, 2008).

From a geographical point of view, the study is important as it analyses not only historical facts, but also socio-environmental facts, such as the degradation of vast areas of native vegetation for the implementation of the hydroelectric plants (Paulo Afonso I, II, III, IV and Apolônio Sales). In addition, of course, to socio-spatial and environmental facts, such as the construction of the city of Paulo Afonso and local and regional development (ANEEL, 2012).

Finally, it should be emphasised that this work has greater academic relevance because of the city's importance to the northeast, its strategic role in the process of industrial acceleration in this region, as a national project to reverse the economic and social differences between the north and south of the country, and above all, the neglect of memory preservation in the city (VASCONCELOS, 2013).

CHAPTER 2

RESEARCH OBJECTIVES

2.1 General

To analyse the socio-environmental degradation caused by the Paulo Afonso hydroelectric complex in Bahia, Brazil, through the socio-spatial consequences caused by the construction of the energy project.

2.2 Specific

- ❖ Understand the historical construction of the Paulo Afonso hydroelectric complex;
- ❖ To identify the socio-environmental changes caused by the hydroelectric complex in the municipality of Paulo Afonso;
- ❖ To evaluate the socio-spatial contribution made by the hydroelectric complex in the city of Paulo Afonso.

CHAPTER 3

THEORETICAL BACKGROUND

3.1 Definition of Hydroelectric

Hydroelectric power is generated by harnessing the flow of water in a plant in which the civil works, which involve both the construction and the diversion of the river and the formation of the reservoir, are just as important, if not more so, than the equipment installed. For this reason, unlike thermoelectric plants (whose installations are simpler), the construction of a hydroelectric plant requires the hiring of the so-called heavy construction industry (ANEEL, 2012).

It's worth pointing out that man has been using hydraulic motors since before Christ, and the first really practical ones were water wheels. Although they were extremely simple and easy to build, they met the demands made of them for centuries. However, as they were used for low drops of less than 6 metres, and also due to their low speed and power, they lost ground as the Industrial Age progressed, reducing them to very special cases (JUNIOR GRAMULIA, 2009).

According to Souza (2011), the main variables used to classify a hydroelectric plant are: height of the waterfall, flow rate, installed capacity or power, type of turbine used, location, type of dam and reservoir. These are all interdependent factors. Thus, the height of the waterfall and the flow rate depend on the construction site and will determine the installed capacity - which in turn determines the type of turbine, dam and reservoir.

Barbosa (2010) says that there are two types of reservoirs: accumulation and run-of-river. The former, generally located at the headwaters of rivers, in places with high waterfalls, due to their large size, allow for the accumulation of a large amount of water and act as stocks to be used in periods of drought. In addition, as they are located upstream of the other hydroelectric plants, they regulate the flow of water that will flow into them, so as to allow the integrated operation of the group of plants. Run-of-river plants generate energy from the flow of water in the river, i.e. from the flow with minimal or no accumulation of water resources.

According to Hambus (2010), a hydroelectric plant is an installation where the potential energy of water gravity is transformed first into mechanical and then electrical energy. It may or may not have an accumulation reservoir, depending on whether the amount of water available varies greatly throughout the year. When there is no reservoir or it is not used for accumulation or flow regularisation, the hydroelectric plant is said to be run-of-river.

The activities or undertakings that potentially modify the environment, listed in CONAMA Resolution 237/97, are subject to environmental licensing, which is defined as:

> **"the administrative procedure by which the competent environmental body** licences the location, installation, expansion and operation of undertakings and activities that use environmental resources, considered to be effectively or potentially polluting or those that, in any form, may cause environmental degradation, considering the legal and regulatory provisions and technical standards applicable to the case" (BRASIL, 1997, CONAMA RESOLUTION No. 237/1997, Article 1).

Pompeu (2006) states that electricity can be obtained from hydroelectric plants, using the energy potential of water, but it can also be produced in wind power plants, harnessing the energy of the wind; in thermoelectric plants, which work by burning some fuel, whether from clean sources or not; through thermosolar plants, which generate electricity from the sun's energy; by thermonuclear plants, which use some nuclear fuel for their operation; among other industrial plants.

According to Salles (1993, p.685), in geographical terms, a river "is any course of fresh or fresh water, larger than a stream, which drains its waters into a defined channel". Schaeffer (1986) states that in hydrological terms, a river corresponds to an open system, with a continuous flow from source to mouth, being considered as a sequence of ecosystems, where the current is the predominant vector that displaces the effects spatially. The author also points out that the water balance and the energy coming from the basin cover (relief, vegetation, soils, etc.) are determinants of the flow and typology of a river.

For his part, Sevá Filho (2004) says that the construction phase of the plant is long, and can take up to 10 years depending on the size of the plant, and includes other activities in parallel, such as population resettlement and the implementation of roads and other infrastructure aspects. The impacts of this phase are many and varied, ranging from those intrinsic to large-scale construction to changes in the water regime and the large number of workers who move into the region during the construction years, which creates pressure on the towns and municipalities neighbouring the site, increasing demand for goods and services, raising property prices and generating local inflation.

According to ANEEL (2005), hydroelectric power is of fundamental importance to the development of today's societies. They can be converted to generate light, power to move engines, industries, run various electrical and electronic products, among a huge variety of purposes.

Moret and Ferreira (2008) define a hydroelectric plant as a set of works and equipment whose purpose is to generate electricity by harnessing the existing hydraulic potential of a river. In other words, the hydraulic potential is provided by the gradients along the course of the river and the hydraulic flow, Figure 1.

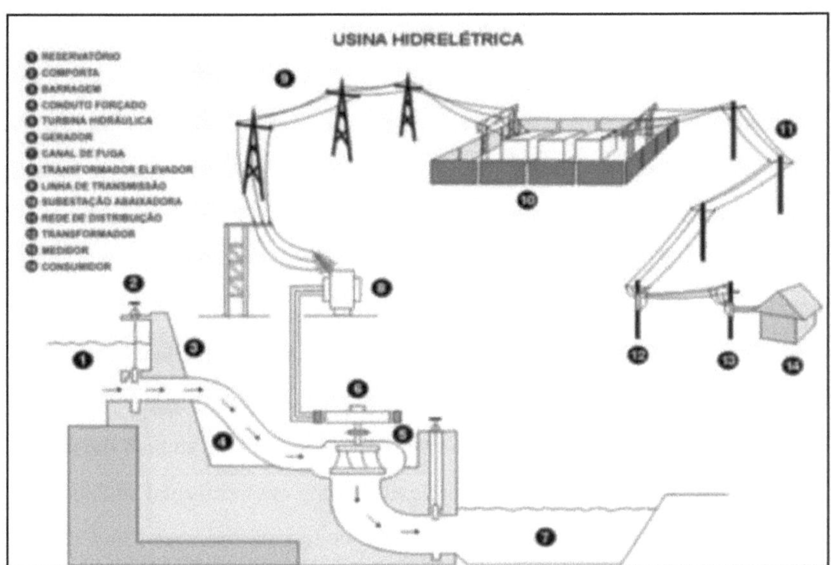

Figure 1. Schematic representation of a hydroelectric power station.
Source: FURNAS (2007).

Finally, according to Lemos Junior (2010), a hydroelectric power station is an architectural complex, a set of works and equipment, the purpose of which is to produce electricity by harnessing the hydraulic potential of a river. According to the author, a hydroelectric power station uses a dam to accumulate water in order to ensure a hydraulic potential that will be used to turn a turbine. This, in turn, mechanically coupled to a generator, will convert the potential hydraulic energy into electrical energy which, through specific equipment such as cables, transformers, towers, etc., will be responsible for providing access to the energy for human use.

3.2 Understanding social, spatial and environmental degradation

The planet's population is totally dependent on its ecosystems and the services they provide, including food, water, disease management, climate regulation, spiritual fulfilment and aesthetic appreciation (JORDÃO, 2009). Natural systems perform vital functions and provide goods and services to human beings, enabling the continuity and maintenance of other species (CONSTANZA et al., 1997).

Around 60% (15 services) of the 24 (100%) surveyed of the services provided by ecosystems examined during the Millennium Ecosystem Assessment have been degraded or used unsustainably, including pure water, capture fisheries, air and water purification, local and regional climate regulation, natural threats and epidemics. Many ecosystem services have deteriorated as a result of actions aimed at intensifying the supply of other services, such as food. In general, these mediations either transfer the costs of degradation from one group of people to another or pass on the costs to future generations (MILLENNIUM ECOSYSTEM ASSESSMENT, 2005).

The intensification of global environmental damage resulting from production activities in industrial capitalist society has led to a considerable advance in scientific research into this issue, forming an interdisciplinary field of research involving scholars from both the natural and social sciences. At the same time, the complexity of the environmental problem has highlighted the serious limitations of the disciplinary studies of the social and natural sciences, carried out to diagnose the root of problems, their prevention and even the formulation of policies to halt or reverse environmental degradation (GARCIA,1994).

This broad dimension of socio-environmental problems also implies the need to promote the participation of the scientific community in international programmes, while at the same time demanding that national policies be solidified. From this perspective, research has been encouraged into the fundamental mechanisms that govern the evolution of the environment and which should be taken as the basis for defining regulatory actions capable of allowing this evolution to be controlled. Thus, according to Jollivet and Pavè (1997), the emphasis should be on understanding the environmental transformations linked to human activities, as well as identifying the possible threats of a global and local nature, related to societies, and finally, proposing an alternative development, generated from new technological, socio-economic and political solutions.

Therefore, the theoretical discussion of socio-environmental value is placed within the vision of integrated heritage conservation, whose approach focuses on natural and cultural assets as a single object, subject to the set of principles, guidelines and actions that make up the heritage conservation management system. The principles and foundations of integrated conservation were established as a heritage management process based on heritage charters, recognised internationally by UNESCO through ICOMOS (International Council on Monuments and Sites) and IUCN (International Union for Conservation of Nature and Natural Resources). As a theoretical-methodological approach, the discussion and implementation of guidelines and actions aimed at protecting the world's heritage has been developing since the 1960s. Initially, this approach focused on the categories of cultural goods in their strictest sense, works of art and historical monuments, and later expanded to include the environment of which they are a part (COELHO; CUNHA, 2003).

This concern is part of the discussion about the evaluation of heritage assets at international, state and even municipal level, given the lack or inadequacy of the theoretical-methodological and operational instruments currently known or in force. The problem of periodically assessing the state of conservation and the permanence of the values of natural and cultural assets has prompted managers and scholars in the field to discuss operational instruments for monitoring the conservation of these assets (LOUREIRO, 2003; HOGAN, 2000).

Nevertheless, the evolution of environmental policy in Brazil took place at the same time as the international process, with national milestones being established after the Stockholm Conference (1972), with the creation of SEMA - the Special Secretariat for the Environment at federal level, and some state environmental bodies. Thus, the 1980s saw the institutionalisation of water resources management, with five ministries responsible for this, all coordinated by the MME. Discussions and new environmental studies point to the limited availability of water resources, increasing the importance of considering water as an economic asset. The need then arises for mechanisms that make it possible to control and monitor works of common interest in order to allow the continued use of water in a sustainable manner (ROMERA; SILVA, 2004).

Rosa et al. (1994) state that the use of water resources in Brazil has resulted in the installation of watertight sectors with independent management. Environmental management must take into account that environmental protection is in the public interest, being a citizen's right and the state's responsibility towards present and future generations, and there is also a legal obligation in this regard. Environmental protection encompasses the three levels of government, federal, state and municipal, all of which have the autonomy to make their own laws. The real problem is not the insufficiency or poor quality of the laws, but their application, as they conflict with particular interests.

According to the Millennium Ecosystem Assessment (2005), the understanding of natural assets lies in the concept of nature and the content of attitudes towards human dominance over it. The relationship between man and nature is understood in the light of the processes of human endeavour over time and the subsequent result of environmental preservationist movements that have changed the direction of understanding the relationship between man and nature. From this perspective, the object of study is placed at the centre of discussions between anthropocentrism, biocentrism and ecocentrism. It is understood that current approaches converge towards an ecosystemic view of the environment.

This discusses the concept of nature and how it has been studied as heritage, its natural significance and the values attributed to it.

For Fahrig (2003), an understanding of nature is based on an understanding of the process of relationships established between man and his natural environment throughout his historical and cultural trajectory on the earth's surface. The theoretical discussion of nature's values is based on the concept of natural significance contained in the assumptions of the Australia/IUCN Natural Heritage Charter, 1996. The principle of existence value lies in the life of living organisms, ecosystems and terrestrial processes, which have values that go beyond social, economic and cultural values. The values of nature preached by the Natural

Heritage Charter fall within the ecocentric vision, which translates as the way in which man relates to nature while respecting its biotechnological cycles.

Brito (2001) emphasises that the value of nature is investigated based on the principle of natural significance and its set of values seen as fundamental elements for contributing to the formulation of assessment, control and monitoring instruments for the integrated conservation management system. The idea is therefore to build a theoretical-methodological basis applicable to the process of developing systems of indicators for the conservation of natural and cultural assets that will make it possible to assess the transformations and changes they have undergone over time. This will enable corrective or mitigating measures to be taken in relation to the state of the heritage asset, the pressures to which it is subjected and the responses given by organised society in favour of this asset.

For his part, Morsello (2001) emphasises that, when viewed in the light of the Sustainable Spatial Development approach, this problem takes on new contours, since the sustainability of a given territory is the result of socio-environmental construction, in which the rules and principles guiding access to and use of natural and cultural heritage are (re)interpreted by local actors. These actors also assume responsibility for action strategies relating to development processes.

However, Andion et al. (2007) point out that socio-economic and socio-environmental dynamics are interpreted based on a vision of the interdependence and coevolution of the relationship between the dimensions of nature and culture. This makes it possible to reconcile an empirical and inductive approach, which favours the analysis of social dynamics in development processes, with a propositional and prospective approach, which aims to assess the consequences of social practices, given the urgency of making new styles of development operational.

3.3 Understanding socio-spatial consequences

Socio-spatial segregation, considered to be an integral and fundamental part of understanding the production of urban space, as discussed in the work by Vasconcelos (2013) and Carlos et al. (2013), has been the growing subject of research in various areas of the social sciences. This urban phenomenon, which manifests itself both through restricted access to urban qualities, as Rolnik (2008) suggests, and through unequal exposure to environmental impacts, as Acselrad (2007) points out, is still far from being exhausted (CORRÊA, 2013).

Netto (2013), by investigating urban phenomena using the socio-spatial theory[2] , was able to provide "empirical support in a field of practice marked by normative approaches largely based on inferences that have never been verified". For the author, this phenomenon presents itself as a novelty in socio-spatial approaches to the theoretical scenario of urban studies and he believes that the application of this theory is capable of translating more clearly the interference that the model of spatial organisation has on social life (CORRÊA, 1989).

Pereira et al. (2011) highlight the great interest in dealing with this problem using the Space Syntax approach, also known as Syntactic Analysis of Space or Social Logic of Space Theory, created in the 1970s by researchers Bill Hillier and Juliene Hanson at University College London (UCL/London - England), with collaborators from various countries, including Brazilian researchers.

According to Spósito (2012), the socio-spatial consequences are part of the process of accentuating forms of segregated appropriation of space, as the models identified in areas of expansion have shaped an increasingly discontinuous and fragmented urban morphology. In this sense, urban transformations have contributed to profound changes in urban configurations, intensifying the phenomenon of socio-spatial segregation in Brazilian cities, whether metropolitan or not (HALL, 2006).

Although recognising the diversity of socio-spatial perceptions and interests is at the root of thinking about territorial development, it helps to explain the exclusions, contradictions and conflicts that are inherent to it, despite the fact that numerous normative formulations insist on valuing the positive aspects of local life, such as the synergy, solidarity and cooperation of the social actors, it is around social exclusions and conflicts that an understanding of the blockages to development dynamics must be sought, which "unfortunately", values, attitudes and behaviours maintain between themselves and with

---

[2] Netto's (2013) analyses show that space is a formation broken down into small nuclei rather than an articulated socio-spatial system. This contribution is based on a critical geographic approach, with the aim of sharpening the critical and investigative spirit of researchers who intend to look at these symbolic urban centres of Brazilian socio-spatial formation, which today are involved in a process guided by the dialectic of destructive construction that entangles the world's cultural assets.

economic variables, links that cannot be equated (VIEIRA; COZELLA, 2006).

With regard to socio-spatial factors, Ruwer (2004) listed some important aspects to consider when inserting a hydroelectric project in a given region, as shown in Table 1. It is important that regional issues are included right from the planning stage, so that the project can mitigate or minimise adverse effects, mobilising resources and contributing to the quality of life of the population directly and indirectly affected.

Table 1. Relevant Aspects in the Insertion of a Hydroelectric Enterprise in a Given Area Region.

FACTORS	CONSEQUENCES
AREA OF INFLUENCE (Municipality(ies) affected, with emphasis on those with altered population and economic demand as a result of the development)	Population dynamics: conditions urban settlement, quality of life and changes in the original urban nucleation.
	Economic dynamics: Behaviour of the municipal economy as a result of the loss of natural resources, loss of agricultural production and deactivation of industrial and commercial establishments. Benefits provided by the project, such as the implementation of tourist activities linked to the reservoir.
	Level of employment and income (urban and rural). Absolute and relative share of the increase in revenue brought about by the development in municipal tax collection; emphasis on possible investment policies in basic infrastructure, driven by the increase in tax collection.
	Increase in municipal and state government spending due to the maintenance of services to meet the needs created *by the* insertion of the enterprise.
	Changes in cultural traditions.
	Benefits of energy at local and regional level.
AREA DIRECTLY AFFECTED (Properties, establishments around the reservoir, the surrounding area and resettlement areas)	Economic activities and infrastructure: land use and occupation; water uses in the water resource harnessed by the plant; primary productive activity; social relations of production; tourist activities and recreation and leisure.
	Population aspects and quality of life: significant changes in numbers, housing standards and occupancy conditions.

Source: Adapted from Ruwer (2004).

Contrary to the thinking of the Chicago School[3] , Vieira and Melazzo (2003) state that the concept of socio-spatial segregation is heavily influenced by Marxist thought and identify Jean Lojkine, Manuel Castells and Henri Lefebvre as the main scholars of this thought. However, for Lefebvre, the organisation of urban space in capitalist society is a form of social

---

[3] Considering space as a broad and complex "social laboratory", sociological research has been marked by the systematic use of empirical methods to collect data and information on urban conditions and ways of life. In other words, the central question is to what extent deviant behaviour (for example, various forms of crime) are products of the social environment in which the individual is inserted (HARVEY, 1996).

production based on three important factors. The first refers to urban space being treated as a commodity, the second is a result of the first, which produces unequal access to urban spaces determined according to social class, and finally, those who believe that, as a result of the first two conditioning factors, the appropriation of space ends up taking on an ideological treatment.

Emphasising analyses and reflections on cities of this size, Maricato (2006) comments that the appropriation of urban land in the expansion areas of these cities has reconfigured the pattern of socio-spatial segregation in such a way that this phenomenon requires a better understanding of its forms and spatialisation. Thus, in view of the above, an important issue is to identify the spatial configuration of cities, which can increase the pattern of socio-spatial segregation (HARVEY, 1993).

In line with Medeiros (2006), the growing impasses experienced in urban spaces have motivated the search for new methodologies and new scales to understand the socio-spatial dynamics of cities. Generally speaking, the analysis of socio-spatial consequences is based on reading disjointed spatial units, such as the centre-periphery relationship, which Villaça (2011) believes is restricted to description. On the other hand, Space Syntax is part of a current that supports a more precise analysis of the urban configuration based on street segments and has the potential to assess the relationship between segregation and its totality with the urban structure which, according to Villaça (2011), without this broader relationship these studies are incomplete.

The topic of socio-spatial impact is nothing new, as it has already been widely explored in academic literature. However, it should be noted that most studies related to urban phenomena have favoured the same scale of approach, i.e. the metropolitan scale, such as the work of Marques (2014), Gómez Sandoval (2011), Luco and Vignoli (2003), Villaça (2001), Caldeira (2000), Ribeiro (1997), Bonduki and Rolnik (1982) and Kowarick (1979), among others. For this reason, it can be said that, although not exhausted, research on this subject is consolidated when it comes to large urban centres.

However, there are already several studies that offer some possibilities for interpreting this phenomenon in medium-sized cities, such as those by Spósito (2013; 2012), Sobarzo (2006; 1999), Bellet Sanfeliu and Llop Torné (2004), Carvalho (2003), and Melazzo (1993), although as far as the study of socio-spatial segregation in medium-sized cities is concerned, they are still under development.

CHAPTER 4

RESEARCH METHODOLOGY

With regard to the fundamental role that the choice of topic plays in scientific research, Macorni and Lakatos (2009) present it as the first step in constructing the scientific object. According to the authors, the choice of topic means recognising both the importance of a particular process among a wide range of others, as well as its scientific relevance, which in fact justifies the research.

As part of this discussion, Gil (2009) also points out that the topic chosen for research is difficult, so to speak, and sometimes tempting. In short, at least in the humanities and social sciences, the subjects we choose have to do with our desires, which explains why they attract and frighten us at the same time. Thus, part of the criticism of the crisis in the way theses are done in the humanities and social sciences is the mechanical relationship that theories have assumed in understanding academic production, which highlight and relate the occurrence of this process to the loss of "the anguish of thinking" and the "libido to know", consequently tending not to advance into the unknown continents of scientific knowledge (FONSECA, 2007).

It is precisely in this tenuous issue that intersects contact with a bibliographical framework that is already known, but which at the same time presents limitations for understanding the emerging processes proposed and which, consequently, require geographic science to discover a "New Continent", that we advocate, alongside Costa (2006), the need to take the theoretical contributions already made as entrepôts to advance "into the sea" and be able not only to propose them, but also (perhaps) to intend them.

Therefore, from a critical point of view, we have taken care to search the academic literature and treat it with special care to avoid falling into the problems that we have discussed and analysed so much theoretically and that, like the territory in Raffestin's conception[4] (MARCONI; LAKATOS, 2009; BAPTISTA; CUNHA; 2009; RAUEN, 2002), the bibliographical references become "the prison that man builds for himself".

---

[4] For Raffestin (1993), it is power relations, through different actors, who appropriate space to form territories, imprinting their relational characteristics on them according to their objectives, which can be influenced economically, politically, culturally and even by the natural environment.

3.4 Choice of Method

The method marks out a route chosen from among other possible routes. However, it is not always the case that the researcher is aware of all the aspects involved in this journey; this does not mean that they are not adopting a method. However, in this case, there are many risks of not proceeding carefully and coherently with the theoretical premises that guide their thinking (MARCONI; LAKATOS, 2009).

In other words, the method is not just any path among others, but a safe path, an access route that allows the social issues proposed in a given study to be interpreted as coherently and correctly as possible, within the perspective embraced by the researcher (BAPTISTA; CUNHA, 2009).

Therefore, the object of the methodology is to study the explanatory possibilities of the different methods, situating the peculiarities of each one, their differences, divergences, as well as the aspects they have in common (GIL, 2009).

Thus, in accordance with Fonseca (2007), with regard to one of the characteristics of research, we can frame the analysis of the research problem as one of the first steps of the investigation, which according to Gil (2009); Marconi and Lakatos (2009); Baptista and Cunha (2009), is the technical, systematic and exact exploration, where the researcher is based on studies already carried out by theorists, in order to be sure of the method to be worked on, evaluating the correct design. Similarly, Silveira (2004) calls this procedure a literature review.

3.5 Research Approach

More than defining the method, for this research we are interested in the basic characteristics of qualitative research. Without pretending to exhaust them, it can be said that they include, according to Flick (2009) and Cassel and Symon (1994, p. 127 - 129): a) a focus on interpretation rather than quantification: generally, the qualitative researcher is interested in the participants' own interpretation of the situation under study;

      b) emphasis on subjectivity rather than objectivity: it is accepted that the pursuit of objectivity is somewhat inappropriate, since the focus of interest is precisely the perspective of the participants;

c) flexibility in the process of conducting the research: the researcher works with complex situations that do not allow for the exact and a priori definition of the paths that the research will follow;

d) process-orientated rather than result-orientated: the emphasis is on understanding rather than a predetermined objective, as in quantitative research;

e) concern with context, in the sense that people's behaviour and the situation are closely linked in shaping experience;

f) recognising the impact of the research process on the research situation: it is accepted that the researcher has an influence on the research situation and is also influenced by it.

g) data is preferably collected in the contexts in which the phenomena are constructed;

h) data analysis is preferably carried out during the data collection process;

i) the studies are descriptive, focusing on understanding in the light of the meanings of the subjects themselves and other references;

j) the theory is constructed by analysing empirical data and then refined by reading other authors;

k) the interaction between researcher and researched is fundamental, which is why the researcher is required to perfect various skills, especially in communication techniques;

l) the integration of qualitative data with quantitative data is not denied, but rather the complementarity of these two models is encouraged.

In view of the above, this study will be conducted using the qualitative approach, in terms of research strategy, because, according to Minayo (2003), this tactic is the path of thought to be followed, because it occupies a central place in the theory, basically dealing with the set of techniques to be adopted in the construction of reality.

In this way, qualitative research is an activity of science that is concerned with the social sciences at a level of reality that cannot be quantified, since it works with the universe of beliefs, values, meanings and other deep constructs of relationships that cannot be reduced to the operationalisation of variables. However, quantitative research can lead the researcher to choose a particular problem to be analysed in all its complexity using qualitative methods and techniques, and vice versa.

Therefore, the integration of qualitative data with quantitative data is not denied, but rather the complementarity of these two models is encouraged (FLICK, 2005; ROSENTAL; FRÉMONTIER-MURPHY, 2001). Since the process is the main focus of this approach, it is not the result or the product, but rather the analysis of the data collected intuitively and inductively by the researcher, and the study is descriptive, focusing on understanding in the light of the meanings of the subjects themselves and other references (FACHIN, 2003; BRANDÃO, 2003). The data is preferably collected in the context in which the phenomenon is constructed (MARCONI; LAKATOS, 2002).

However, it should be emphasised that the study proposed in this document will be of the qualitative research type, as it allows for a greater understanding of the data collected in the laboratory (Computer Room and Libraries) and in the field (in loco).

3.6 Content Analysis

Content analysis arose from the need to systematise the rules and the interest in extending the applications of the technique to different contexts and the emergence of new problems in the methodological field that affect research and data analysis. According to Bardin (2011), the primary function of content analysis is critical unveiling. Since content analysis studies aim to focus on different sources of data (BARDIN, 2011).

The first phase is concerned with objectivity in the analyses, overcoming uncertainties and enriching the readings. Content analysis is therefore defined as an empirical method. According to Bardin (2011), content analysis is a set of methodological tools that is constantly being improved and applied to extremely diverse discourses (contents and continents).

After the first, Bardin (2011) and Fonseca (2007) point out the analytical description, presenting the probable applications of content analysis as a method of categorisation that allows the components of the meaning of the message to be classified in a kind of drawer. According to the authors, content analysis is an analysis of meanings, as it deals with an objective, systematic, qualitative and quantitative description of the content extracted from communications and its respective interpretation.

Although their differences are marked, content analysis is concerned with knowing what lies behind the meaning of words, since Bardin (2011) mentions some techniques and procedures of content analysis, such as document analysis, as a way of condensing information for consultation and storage (Chart 2).

Table 2. Distinctions between Documentary Analysis and Content Analysis.

Documentary Analysis	Content Analysis
Focuses on documents; Classification - Indexing; Objective: condensed representation of information for consultation and storage.	It focuses on messages (communications); Categorical-thematic (it is only one of the possibilities for analysis); Objective: manipulation of messages to confirm indicators that allow inferences to be made about a reality other than that of the message.

Source: Adapted from Bardin's study (2011).

Therefore, data collection through the observation technique seeks to obtain information, using the senses in the process of reaching certain aspects of reality, which at first sight are incomprehensible. It is an anthropological research tool that is a fundamental research technique. This technique helps the researcher to obtain and identify evidence about objectives that individuals are not aware of, but which guide their behaviour. Observation plays an important role as it forces the researcher to establish direct contact with the reality being studied (FACHIN, 2003; MARCONI; LAKATOS, 2002).

The researcher is also free to move around, change the focus of observations or concentrate on unexpected facts, as well as allowing for comparison between the information received from the people being researched and reality itself (RAUEN, 2002).

Finally, as far as practices are concerned, some examples illustrate a safe and objective content analysis, such as interview analysis, which is a specific research method and is classified as closed and open. For this research, the best of the two will be used, i.e. semi-structured interviews (open and closed). Furthermore, we emphasise that the treatment of the results comprises coding and inference.

3.7 Research procedures

The methodological procedures are in line with the guidelines of the Academic Institution, which can be interpreted in four inseparable moments: firstly, a dissertative-argumentative exposition on the stages of the document, i.e. generating theme, study proposal, importance, methodology, agent and actors involved in the study, results obtained.

Next, taking into account the historical aspects and the planning involved in the construction of the hydroelectric project, we searched the bibliography to draw up the Theoretical Framework, in order to better understand the theoretical-practical scope, as well as the unfolding of the categories of analysis.

In a third step, the research will be carried out using a content analysis, in which the data, information and documents pertaining to the area where the hydroelectric complex is located will be evaluated, and studies will be carried out on the categories of analysis pertaining to the subject.

Afterwards, we carried out the practical work, collecting data from visits and observing the area (Paulo Afonso Hydroelectric Complex, municipality of Paulo Afonso, Bahia, Brazil), as well as planning the next activities.

In the fourth and final stage, in parallel with the on-site interventions, a document was drawn up on the study, which consisted of an elaborate interdisciplinary perspective. Finally, the results were analysed through the descriptive memorial of the research, socialising it with the academic and school communities, as well as with civil society, fulfilling the objective of science to universalise the knowledge developed in academia.

CHAPTER 5

CHARACTERISATION OF THE OBJECT OF STUDY

The Paulo Afonso/BA region began to be inhabited by Portuguese bandeirantes at the beginning of the 18th century. In 1725, the sesmeiro Paulo Viveiros Afonso was granted a sesmaria on the left bank of the river, on the Alagoas side, which included the lands of the waterfall, until then known as Sumidouro. Later, in 1913, Delmiro Gouveia, an industrialist and businessman at the time, saw the potential of the region and set up a big, bold project, the first hydroelectric power station in the Northeast, the Angiquinho Power Station (FOLHA SARTANEJA, 2015, p. 9).

It was based on the idea of the pioneer Delmiro Gouveia that the then President of Brazil, Getúlio Vargas, signed the Decree authorising the organisation of CHESF - Companhia Hidrelétrica do São Francisco, which was made official in 1948 with the first General Shareholders' Meeting (MUCCINI; MALTA, 2007).

Nevertheless, the population of Paulo Afonso was formed by a mixture of people who came from various states of the country to work on CHESF's construction projects, and from these pioneers came the new generation that populates the city today. As a relatively young city, Paulo Afonso does not yet have a formed cultural identity, and suffers influences from its neighbouring municipalities. Various cultural projects are being developed, and many spaces have been built for their practice (REIS, 2004).

Around CHESF, what would become the city of Paulo Afonso (Figure 2) was born, until then part of the municipality of Glória/BA. It wasn't until 1958 that the municipality was born, through its political emancipation. Paulo Afonso is still a young city with a great future ahead of it, a city led by the strong hands of a hard-working and cheerful people. Today it is considered one of the best cities in the Northeast, dynamic and with great potential for growth, the true "Oasis of the Sertão" (TEIXEIRA, 2005).

In turn, the local Bahianness emerges with the people during their leisure time. The friendliness, charisma and joy of the pauloafonsino make the place pleasant, and add to the quality of life found there. Nature, which has rewarded the region with beautiful landscapes, plenty of water and energy, is the perfect setting to establish the city of the future (FERRAZ,

1999).

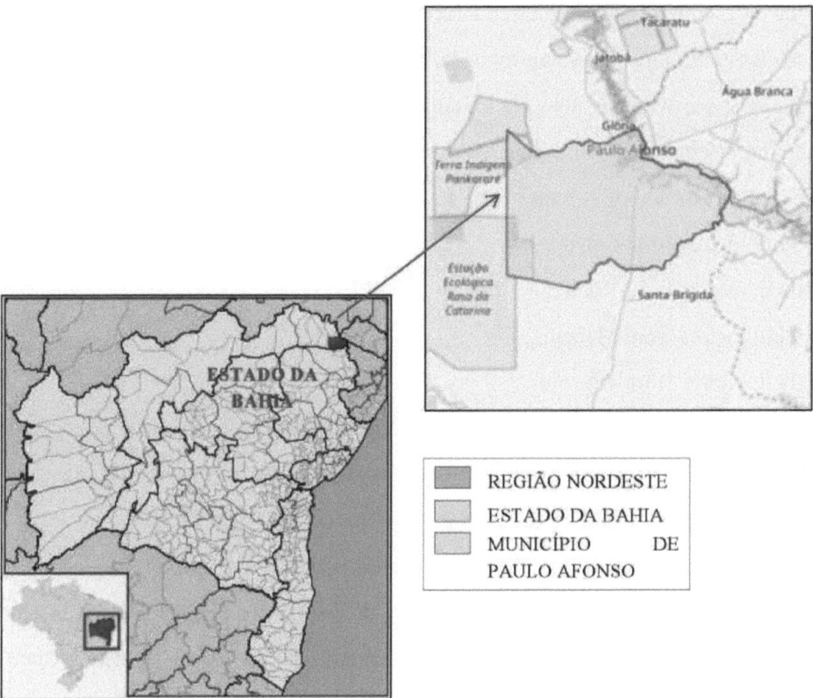

Figure 2 - Location of the study area.

Source: Adapted by the author from IBGE data (2018).

5.1 Socio-environmental aspects of the study area

The territorial area of Paulo Afonso (Figure 2) belongs to homogeneous micro-region 147, Sertão de Paulo Afonso, occupying a territorial area of 1,545.192 square kilometres (km^2) (Table 3) (IBGE, 2017).

Table 3. Data from the municipality of Paulo Afonso/BA, Brazil.

Estimated population 2017	**120,706 thousand inhabitants**
Population 2010	108,396 thousand inhabitants
Area of the territorial unit 2016 (km)^2	1.545,192
Population density 2010 (inhabitants/km)^2	68,62
Municipality code	2924009
Gentile	Paul-Afonsino
Mayor 2017	LUIZ BARBOSA DE DEUS

Source: IBGE (2018).

The municipality is bordered:

- to the north, with the municipality of Glória, from which it was separated in 1958;
- to the south, with the municipalities of Jeremoabo and Santa Brígida and the state

of Sergipe;

- to the east, with the São Francisco River and the state of Alagoas;
- to the west, with the municipality of Rodelas.

The São Francisco River is also the landmark that separates the states of Bahia and Alagoas, Bahia and Pernambuco and Alagoas and Sergipe. The seat of the Municipality of Paulo Afonso is at an altitude of 243 metres and is remote (IBGE, 2017):

- 460 kilometres from Salvador;
- 480 kilometres from Recife;
- 380 kilometres from Maceió;
- 280 kilometres from Aracaju.

In terms of population, the municipality of Paulo Afonso/BA has finally passed the 100,000 mark. With the finalisation of the 2010 census (IBGE, 2010), the municipality had 108,396 inhabitants in 2010, 11,897 more than the 2000 census of 96,499 inhabitants. Currently, in 2017, Paulo Afonso has 120,706 inhabitants (Chart 3) (IBGE, 2017).

In the municipalities of Glória and Santa Brígida, the population fell. Glória had 14,559 inhabitants and now has 13,879, i.e. 680 fewer. In Santa Brígida, the drop was even greater: 1,379 inhabitants, from 16,903 to 15,524 in 2017 (IBGE, 2017).

The relief of the municipality of Paulo Afonso is made up of plateaus and depressions, represented by crystalline soil, as well as tablelands made up of the sedimentary layers of the Tucano-Jatobá Basin. Physical landmarks include: Cachoeira de Paulo Afonso, with several falls: Croatá, Véu da Noiva, the canyon of the São Francisco River, which begins at Cachoeira de Paulo Afonso, carved into pre-Cambian, Archaeozoic migmatous soils and the Serra do Umbuzeiro, in Povoado Riacho. Another point in the municipality is part of the Raso da Catarina, which stretches for 6,000 kilometres. About 40 kilometres from the centre of Paulo Afonso is the place known as Baixa do Chico, on the border with Brejo do Burgo, in the municipality of Glória. There is a dry canyon about 12 kilometres long, with curious sandstone formations (IBGE, 2017).

Located in the Sertão region of the Northeast, Paulo Afonso has a semi-arid BSh *(Koeppen)* climate, with average annual rainfall of between 500 and 600 millilitres. Its average temperature is high, around 30°, reaching 40° in the hottest periods

(December/January). The hottest months are from October to January and July is the coldest with a temperature of around 22° (IBGE, 2017).

In terms of vegetation, the caatinga is the predominant formation in the region, influenced by the BSh climate. As well as trees and low shrubs with twisted branches, there are many species that store water in their stems and roots, such as mandacarus cacti, facheiros, xique-xique, coroas de frade (and the umbuzeiro). The Croatá is very common in the region and also stores water, being the salvation of hunters and residents of the Raso da Catarina region on many occasions. Paulo Afonso's soil is siliceous, almost humus-free, poor in nitrogen[5] and with a regular potassium and calcium content. Plant species such as: umbuzeiro, baraúna, jatobá, caraibeiras, bromeliáceas and cactáceas and others typical of the caatinga are found in the region (IBGE, 2017).

3.8 Description of the object of study

The Paulo Afonso hydroelectric complex, the subject of this study, is made up of the Paulo Afonso I, II, III, IV and Apolônio Sales (Moxotó) plants. The complex produces 4 million, 279 thousand and 600 kW (Table 4). Energy is generated from the force of the Paulo Afonso waterfall, a natural drop of 80 metres in the São Francisco River (CHESF, 2017).

Table 4. Electricity generation capacity of the Paulo Afonso complex.

Hydroelectric	Date of Operation	Units	Total power/ kW
Paulo Afonso I	1954	3	180.001
Paulo Afonso II	1961	6	443.000
Paulo Afonso III	1971	4	794.200
Apolônio Sales (Moxotó)	1977	4	400.000
Paulo Afonso IV	1979	6	2.462.400
Total	2017	19	4. 279. 600

Source: Chesf (2018).

---

[5] Nitrogen is a chemical element with the symbol N, atomic number 7 and atomic mass 14.00674 u (7 protons and 7 neutrons), represented in group (or family) 15 (formerly VA) of the periodic table.

CHAPTER 6

RESULTS AND DISCUSSION

Emphasising the spatial cut-off, the research will only look at the hydroelectric plants located in the municipality of Paulo Afonso/BA that make up the Paulo Afonso Hydroelectric Complex (PA), i.e. PA I, PA II, PA III, PA IV and Apolônio Sales (CHESF, 2017).

6.1 Historical aspects of the Paulo Afonso Hydroelectric Power Plants

In Brazil, the first hydroelectric projects took place in the states of Minas Gerais and São Paulo at the end of the 19th century. Subsequently, some attempts were made to invest in hydroelectric power generation and, in the first decade of the 20th century, this type of energy surpassed the production of thermoelectric plants (ANNEL, 2012).

It is worth noting that investments in the generation, transmission and use of electricity were made by foreign groups through financial and technological resources (ROSA, 2007). The state's interventionist measures from the 1930s onwards culminated in the signing of the Water Code in July 1934 as a guideline for water and electricity concessions (OLIVEIRA, 2005).

The Second World War, which broke out in September 1939, aggravated the energy crisis. The difficulties of importing coal and oil derivatives, on which the nation depended heavily, were compounded by the new energy needs generated by the industries' efforts to meet the war needs of the Allied nations (SANCHES, 2011).

The new demands could not be met, both because the electricity sector was struggling financially to expand its generating capacity and because the hydroelectric plants were producing as much as the hydric conditions of unstabilised rivers allowed. On the other hand, the countries that manufactured hydroelectric components dedicated themselves exclusively to the production of war material, interrupting the export of generators, which were necessary for the expansion of the sector (ROSS, 1999).

As a result, at the end of the Second World War in 1945, the Brazilian government was very concerned about producing electricity using the waters of the São Francisco River at the Paulo Afonso waterfall, as Delmiro Gouveia from Ceará had done in 1913 when he built Angiquinho (MUCCINI; MALTA, 2007).

For this reason, Muller (1995) states that the solution to the energy crisis during the Second World War was rationing; however, the problem had been intensifying since the Vargas coup of 1937. With the creation of the Estado Novo and the New Constitution, new hydroelectric projects involving foreign companies were banned (SANTOS, 2008). From this period onwards, the participation of state and federal governments as shareholders in generating and distribution companies intensified, as well as investing in their own companies (MIELNIK; NEVES, 1988). In view of this, the most important initiatives for hydroelectric exploitation of the São Francisco River in the period prior to the creation of CHESF were: Angiquinho (Delmiro Gouveia), Itaparica Hydroelectric Plant, formerly Petrolândia (PE) and the Pilot Plant (MEMORIAL CHESF, 1998).

Brazil's first hydroelectric power station, Angiquinho, located on the Alagoas bank of the São Francisco River and with 1,500 HP (1,102 KW) of power, was the first project to harness the hydraulic potential of the Paulo Afonso waterfall, as well as being one of the first hydroelectric power stations in the north-east of Brazil (Figure 3). Inaugurated on 23 January 1913 by industrialist Delmiro Gouveia, the small plant was intended to power the machines of a thread and yarn industry, Companhia Agro Fabril Mercantil, located in the Alagoas municipality of Pedra (NASCIMENTOS, 1998; SANT'ANA, 1996; MELLO, 1993).

Figure 3. Angiquinho, the first hydroelectric power station in the Northeast and Brazil.

The energy produced was also used to supply electricity to the factory's workers' village. Delmiro even started building the second stage of his plant, near Furna dos Morcegos, which was interrupted when he died on 10 October 1917 (FOLHA SERTANEJA, 2015, p. 9).

The second development, the Itaparica Hydroelectric Power Station, was located at the Itaparica waterfall on the borders of the states of Bahia and Pernambuco, near the old town of Petrolândia - PE. Its construction can be explained by the fact that, during the 1940s, there were plans to make full use of the São Francisco River in successive stages from this waterfall (JUCÁ, 1982).

However, the third project was approved by the Ministry of Agriculture in its explanatory memorandum No. 598 of 23 May 1944, which authorised the installation of a 2,500 KW pilot plant in Paulo Afonso (NASCIMENTO, 1998).

6.1.1 <u>Characterisation of the regulatory bases of Companhia Hidrelétrica do São Francisco</u>

When the São Francisco Hydroelectric Company (CHESF) was founded, the so-called Pilot Plant was already under construction in Paulo Afonso. It was a plant that would supply energy to the towns near the development. However, when the company started building Paulo Afonso I, it had to build the workers' houses that needed power, so the pilot plant was finished and passed to CHESF, to power the street lighting, the construction of the houses and to move the machines (OLIVEIRA, 2001).

However, one of the first steps towards consolidating CHESF as a company in the national hydroelectric sector took place on 3 October 1945, when President Getúlio Vargas signed Decree-Law No. 8.031, authorising the organisation of CHESF by the Ministry of Agriculture; Decree-Law No. 8.032, granting the Ministry of Finance a special credit of 200 million cruzeiros for the subscription of the company's shares and Decree-Law No. 19.706, which granted CHESF a licence, for a period of fifty years, to harness the hydraulic energy of the São Francisco River between Juazeiro (BA) and Piranhas (AL); supply public service concessionaires and distribute electricity directly to a large part of the Northeast. The initial area, delimited by the latter decree, was a circle with a radius of 450 kilometres around Paulo

Afonso, comprising 347 municipalities in the states of Piauí, Ceará, Rio Grande do Norte, Paraíba, Pernambuco, Alagoas, Sergipe and Bahia, totalling 516,650 square kilometres (CHESF, 2002).

According to Muccini and Malta (2007), contrary to popular belief, five hydroelectric power stations had to be built between 1949 and 1985, forming the largest area of hydroelectric power stations in Brazil, the Paulo Afonso hydroelectric complex, with an enormous power generation capacity. In addition, economic growth processes were unleashed in the commerce and services sectors, forming an economic enclave area in the semi-arid region of the Brazilian Northeast.

However, there were many difficulties in building the first power station, many of which were of a technical nature or related to the weakness of the regional infrastructure at the time, which was unable to absorb the structural difficulties of a project the size of Paulo Afonso (HAMBUS, 2012).

With regard to regional road infrastructure difficulties, the biggest problem seems to have been the roads that couldn't support the weight of the large mechanical structures that had to be transported from the ports of Recife and Salvador to the construction site in Paulo Afonso (SANTOS, 2011; 2010).

However, the problems faced generated stories of overcoming, real challenges that had to be overcome. It can be inferred from analysing the documents in the CHESF memorial, as well as newspaper articles from the time, not forgetting academic research on the subject, that the biggest problem in the construction of the hydroelectric plants was transporting the parts for the plant turbines, as these were too heavy for the bridges that couldn't support such structures, forcing the builders to improvise, getting around the difficulties, despite the technical limitations faced at the time (FOLHA SERTANEJA, 2015).

However, the final process that made it possible to dam the river was the construction of steel meshes woven together and placed over the river. Pieces of rock were thrown over this interwoven structure until the point at which the course of the river was completely dammed (TEIXEIRA, 2005).

Paulo Afonso was the first power station in Brazil to be built underground. Due to the fluctuating water levels in the canyon where the discharge channel ends, open-air construction was not recommended. Installation at a very high point, free from flooding, would cause a great loss of kilowatts, while installation at a lower point, to make better use of the waterfall, would require the construction of shields around the plant (MENDONÇA; BRITO, 2007).

Thus, the Paulo Afonso hydroelectric plant (PA I) began to be built in the first quarter of 1949, the control house would be on the surface, and the lift substation would be installed on the Bahian shore, from where two 220 kV transmission trunk lines would run to Recife and Salvador. And from Recife, other lines to João Pessoa, Campina Grande and Maceió (GOMES, 1986).

6.1.2 <u>Paulo Afonso power stations (PA I, PA II, PA III, PA IV and Apolônio Sales)</u>

In 1953, CHESF again negotiated with the federal government to release funds for Paulo Afonso's first expansion plan, which included the installation of the third generating unit, the expansion of the plant's transmission and transformation system, the excavation of the second engine room (later called Paulo Afonso II) and the execution of complementary civil works (SANTOS, 2008; SANTOS, 2002).

The São Francisco diversion works, which had to overcome major technical obstacles due to the depth of the riverbed and the force of its waters, were completed in September 1954 (CHESF, 2017).

The Paulo Afonso I Power Station, Figure 4, built and designed by CHESF, is installed on the São Francisco, the main river in the northeastern region, with a drainage area of $605,171 \text{ km}^2$, a hydrographic basin of around $630,000 \text{ km}^2$, with a length of 3,200 km, from its source in the Serra da Canastra in Minas Gerais, to its mouth in Piaçabuçu/AL and Brejo Grande/SE (CHESF, 2017).

Figure 4: Paulo Afonso I hydroelectric plant.

Source: Chesf (2017).

The Paulo Afonso I plant consists of three generating units powered by Francis turbines[6] , with a unit power of 60,000 kW, Table 5, totalling 180,001 kW (CHESF, 2017).

Table 5. Primary data for the Paulo Afonso I Hydroelectric Plant.

Type of construction	Underground
Designer	CHESF
Building company	CHESF
Home Works	1948
Start Operation	12/1954
Rio	San Francisco
Longitude	38° 16' West
Latitude	9° 22' South
Municipality / State	Paulo Afonso - BA
Installed power	180,001 kW (3 UGs)
Powerhouse length	60,37 m
Powerhouse height	31,0 m
Powerhouse width	15,0 m

Source: CHESF (2018).

The Paulo Afonso II hydroelectric plant, Figure 4, part of the Paulo Afonso Complex, is located in the city of Paulo Afonso, in the state of Bahia. The Paulo Afonso II Plant, built and designed by CHESF.

---

[6]Hydraulic turbine with radial flow from the outside in, designed by Jean-Victor Poncelet around 1820 and perfected by the American engineer James Francis in 1849.

Figure 5: Paulo Afonso II hydroelectric plant.

Source: Chesf (2018)

The Paulo Afonso II plant consists of 6 generating units driven by Francis turbines, 2 units with a unit power of 70,000 kW, 1 unit with a unit power of 75,000 kW and 3 units with a unit power of 76,000 kW, totalling 443,000 Kw, Table 6, (CHESF, 2017).

Table 6. Paulo Afonso II technical data sheet.

Type of construction	Underground
Designer	CHESF
Building company	CHESF
Home Works	1955
Start Operation	1961
Rio	San Francisco
Longitude	38° 16' West
Latitude	9° 22' South
Municipality / State	Paulo Afonso - BA
Installed power	443,000 kW (6 UGs)
Powerhouse length	104,00 m
Powerhouse height	36,87 m
Powerhouse width	18,00 m

Source: Chesf (2017).

The Paulo Afonso III Plant (PA III), Figure 6, built and designed by CHESF between 1967 and 1971, is located on the São Francisco, the main river in the northeastern region, with a drainage area of 605,171 km^2 , a hydrographic basin of around 630,000 km^2 , and a length of 3,200 km, from its source in the Serra da Canastra in Minas Gerais, to its mouth in Piaçabuçu/AL and Brejo Grande/SE, Brazil. The PA III hydroelectric scheme, part of the Paulo Afonso Complex, is located in the city of Paulo Afonso, state of Bahia (CHESF, 2017).

Figure 6 - Image of Paulo Afonso III.

Source: Chesf (2017).

The Paulo Afonso I, Paulo Afonso II and Paulo Afonso III power stations are in the same impoundment, consisting of a reinforced concrete gravity dam, with a maximum height of 20 metres and a total crest length of 4,707 metres.707m, associated with concrete structures such as: 01 (one) Krieger-type spillway, with free discharge, 04 (four) surface spillways, with wagon gates, 01 bottom spillway, 2 sand drains, water intake and underground powerhouse, excavated in solid rock, with a depth of approximately 80m. The PA III plant has 4 generating units driven by Francis turbines, with unit power of 198,550 kW, totalling 794,200 kW, Table 7 (CHESF, 2017).

Table 7. Characterisation of PA III.

Type of construction	Underground
Home Works	1967
Start Operation	1971
Rio	San Francisco
Longitude	38° 16' West
Latitude	9° 22' South
Municipality / State	Paulo Afonso - BA
Installed power	794,200 kW S 4 UGs)
Powerhouse length	127,0 m
Powerhouse height	46,45 m
Powerhouse width	18,50 m

Source: Chesf (2018).

The Apolônio Sales (Moxotó) Plant, Figure 7, is located in the municipality of Delmiro Gouveia/AL, 8 kilometres from the city of Paulo Afonso/BA. Part of the Paulo Afonso Complex, the plant is located about 3 kilometres upstream, before the Delmiro Gouveia dam, so that the water turbocharged by its machines also drives the Paulo Afonso I, II and III plants.

In a second cascade and through a channel dug from its right bank, the Moxotó reservoir supplies the water needed to run the Paulo Afonso IV power station, which is located parallel to it (CHESF, 2017).

Figure 7. Apolônio Sales Hydroelectric Power Station, Alagoas, Brazil.
Source: Chesf (2018).

Apolônio Sales, built and designed by CHESF, is located on the São Francisco, the main river in the northeastern region of Brazil, with a drainage area of 605,171 km^2 , a hydrographic basin of around 630,000 km2 and a length of 3,200 km, from its source in the Serra da Canastra in Minas Gerais to its mouth in Piaçabuçu/AL and Brejo Grande/SE (CHESF, 2017). The Moxotó impoundment consists of a mixed earth-fill dam, with a maximum height of 30 metres and a total crest length of 2,825 metres, associated with concrete structures such as: 01 (one) bottom spillway, 01 (one) spillway with controlled discharge equipped with 20 sector-type gates, with a maximum discharge capacity of 28,000 m3/s and a powerhouse with 04 generating units, driven by Kaplan turbines, each with 100,000 kW, totalling an installed power of 400,000 kW, Table 8. The energy generated is transmitted through a booster substation with six 80 MVA transformers that raise the voltage from 13.8 kV to 230 kV (CHESF, 2017).

Chart 8. Apolônio Sales configuration.

Home Works	**15/ 01/ 71**
Start Operation	04/1977
Rio	San Francisco
Longitude	38° 11'West
Latitude	9° 17' South

Municipality / State	Delmiro Gouveia - Alagoas
Type of construction	External
Installed power	400,000 kW (4 UGs)
Powerhouse length	192,0 m
Powerhouse height	61,00 m
Powerhouse width	22,90 m

Source: Chesf (2017).

The Paulo Afonso IV hydroelectric plant, Figure 8, part of the Paulo Afonso Complex, is located in the city of Paulo Afonso, in the state of Bahia. The plant is installed on the São Francisco, the main river in the northeastern region, with a drainage area of 605,171 km^2 , a hydrographic basin of around 630,000 km^2 , stretching 3,200 km from its source in the Serra da Canastra in Minas Gerais to its mouth in Piaçabuçu/AL and Brejo Grande/SE (CHESF, 2017).

Figure 8. Paulo Afonso IV hydroelectric plant.
Source: Chesf (2017).

This plant receives water from the Moxotó reservoir via a bypass channel. The turbinated water, together with the water turbinated in Paulo Afonso I, II and III, flows through the canyon to the Xingó Dam, Alagoas, Brazil. The Paulo Afonso IV impoundment is made up of dams and dykes with a mixed earth-rockfill section for a total length of 7,430 m and a maximum height of 35.00 m; concrete structures for a total length of 1.053.50m comprising: spillway with 8 crest/controlled gates with a discharge capacity of 10,000 m^3 /s, water intake, underground engine room with 6 generating units each with a nominal capacity of 410,400 kW, Table 9, totalling 2,462,400 kW (CHESF, 2017).

Table 9. Data from Paulo Afonso IV.

Type of construction	Underground
Home Works	1972
Start Operation	1979
Rio	San Francisco
Longitude	38° 16' West
Latitude	9° 22' South
Municipality / State	Paulo Afonso - BA
Installed power	2,462,400 kW (6 UGs)
Powerhouse length	210,00 m
Powerhouse height	52,00 m
Powerhouse width	24,20 m

Source: Chesf (2017).

6.2 The socio-environmental impacts caused by the construction of hydroelectric dams

The milestone in the development of the environmental debate was the World Conference on Environment and Development, organised by the UN (United Nations Organisation) in Stockholm, Sweden, in 1972, followed by the publication of the Brundtland Report, Our Common Future (1987), which discussed the development and economic growth model adopted in recent decades by developed countries and some developing countries. These discussions gave rise to the concept of sustainability[7] , assigning guidelines and principles that go beyond the concept of development based on economic principles, proposing sustainable growth and development[8] in which the use of resources occurs in such a way as to meet the needs of the present without compromising future generations (LOUREIRO, 2012).

Nevertheless, the ecological movement emerged as a warning about the consequences of the advance in industrial technologies, which ended up transforming almost everything, creating an industrial society, bringing serious risks to natural resources and to man, or rather to living beings as a whole (MENEZES; ROCHA, 2010).

---

[7] According to Loureiro (2012) and Marcomim and Silva (2010), the notion of sustainability has a polysemic meaning. The polysemic meaning attributed to the concept ends up not clarifying or contributing to a scientific definition. The notion of sustainability is much more of a political guideline of goals to be achieved than an expression of reality. It has emerged as an agenda, with the aim or ideal of restoring the balance in the relationship between man and nature, especially after the damaging environmental consequences caused by industrialism and its green revolution. In this sense, although many definitions allude to the social, economic and political dimension, they centre their attention on biophysical resources, or start from the preservation of natural resources by relating them to social issues, as if the latter were subordinate to the former.

[8] Sustainable development is defined as development that meets the needs of the present without compromising the ability of future generations to meet their own needs (SANTOS, 2010).

In this sense, there has been investment in hydroelectric potential, as this type of energy would be a way out of the energy crisis, as well as being renewable and clean, because of the energy sources currently being exploited, hydroelectric energy is non-polluting, has no waste and can be reused downstream. However, the fact that it is considered clean energy does not include the construction of hydroelectric dams, which cause profound socio-environmental impacts in order to generate energy (SANTOS, 2008).

Along these lines, according to Pereira et al. (2005) and Ruwer (2004), the implementation of large projects has unfortunately led to experiences in which societies have seen their economic foundations and socio-cultural values suddenly shaken. Although hydroelectric generation is sustainable, some of the regions affected by it have experienced unsustainable regression instead of development.

In the 1980s we were experiencing an environmental crisis. In Brazil, the political changes of the late 1980s led to public demonstrations for advantages and benefits for the populations affected by hydroelectric projects. This was when, in the biological and social sciences applied to the sector, the emphasis on the biotic-anthropic binomial gave way to the political-economic one, in fact, with a delay compared to the countries where this transition took place after the Stockholm Conference (BRASIL, 2004).

These environmental issues gained ground in national and international public opinion, and were reflected in the enactment of the National Environmental Policy Law, Law 6.938 of August 1981 (SANTOS, 2010). Subsequently, Resolution 001/86 defined the basic criteria and general guidelines for the use and implementation of Environmental Impact Assessment. Subsequently, the National Environment Council (CONAMA) issued Conama Resolution 006, of 6 September 1987, which established the criteria for environmental licensing of large-scale works such as power generation (BRASIL, 1987).

The environmental impacts caused by the construction of hydroelectric dams are numerous, according to Sevá (2009). There have been several cases of small lakes bursting and overflowing, and there is also a risk for large dams. This possibility increases as the built structure ages through seepage in the walls and storage capacity is reduced due to silting (MILARÉ, 2006).

Another issue related to the impacts is the rise in the water table[9] in the region. The water often becomes unfit for consumption, damaging the supply to neighbouring populations (MOTA, 2005).

According to Santos (2010), undertakings and/or activities that use environmental resources in such a way as to generate some kind of degradation or pollution must be subject to prior environmental licensing, which must be carried out by the competent environmental body, IBAMA - the Brazilian Institute for the Environment and Renewable Natural Resources, at federal level, and the state and municipal environmental bodies.

In turn, it should be emphasised that every engineering project has an environmental impact to be assessed. Milaré (2006) establishes three ideological operations hidden in the notion of environmental impact. The first appears when analysing the implementation of hydroelectric projects. The project already appears as a fait accompli, unchangeable, and all that is left to do is accept and adapt. The second ideological operation is the one that considers the impacted populations to be part of the environment in which the work will take place.

Consequently, the affected parties are not consulted and only options for minimising negative impacts are seen. There are therefore two entities in confrontation: the state, which is the cause and the agent, and nature, which is patient and only reactive. Engineering becomes social engineering (SCARPIN, 2012).

In short, the state becomes identified as Brazilian "society" in general, and real societies are just objects of the state. The work becomes a natural and inevitable event. As a result, the notion of the environment and environmental impact is incorporated into the state's ideological arsenal, concealing the entire process of political domination (SCARPIN, 2012).

6.3 Socio-spatial consequences of the implementation of the hydroelectric complex

In Brazil, between 1960 and 1990, the construction of hydroelectric power stations caused irreparable impacts on the environment and the local population affected by the

---

[9] The rise in water levels during the formation of reservoirs causes fantastic hydrostatic pressure on the artesian springs located on the banks and bottom of dammed rivers. This process produces degrees of alteration in the entire natural process of feeding and discharging aquifers, including deep ones (RODRIGUES; ADAMI, 2011).

formation of the large lakes. Thus, the construction of hydroelectric power stations in the country has been accompanied by countless socio-spatial conflicts caused by the removal of the affected population and the serious environmental problems resulting from their construction (FOLHA SERTANEJA, 2015).

Thus, in that period, the electricity sector's plans did not include any discussion with society or an action plan that took into account the affected population and environmental and socio-spatial issues, which allowed for the construction of large hydroelectric projects (PEREIRA, 2012).

From the point of view of projects aimed at deeper socio-spatial transformations of the traditional structures that existed in the region, there were few pioneering attempts to exploit the São Francisco River. During the regency period, we highlight the inventory carried out by Hafeld, who from 1852 to 1854 made an inventory of the river's potential. This document describes the physiographic characteristics of the São Francisco River, particularly highlighting, in his own way, the limits and possibilities of using the river for navigation, interspersed with a brief description of the settlements that this naturalist encountered on his excursion (SANTIN; FLORES, 2006).

With regard to the area where the Paulo Afonso Hydroelectric Complex was built, the imposing nature of the Paulo Afonso Waterfall, in the municipality of Paulo Afonso, Bahia, Brazil, stands out, revealing the symbolic grandeur of this waterfall. Also noteworthy are the waterfall's geomorphological characteristics and the wealth of rapids in the area (TEIXEIRA, 2005).

According to Jucá (1982), at the beginning of the 20th century, up until 1910, there were some attempts to requisition the use of the São Paulo River.
Francisco. The author reports at least two projects aimed at implementing regional development through the region's agricultural, energy and industrial potential. The first was by an Englishman, Richard George Reyde, and the second by a Brazilian, Francisco Pinto Brandão, whose applications were rejected on the grounds that the development of the region's potential should be under the auspices of the public authorities.

Despite the hiatus that formed (1917 to 1949), the region continued to be the subject of studies and inventories. All of them pointed to the possibility of harnessing the agricultural and hydroelectric potential of the São Francisco River. Of particular note was the construction of the Paulo Afonso hydroelectric power station, which by the end of the 1940s and beginning of the 1950s had become an indispensable necessity, and was seen by the federal government (Dutra's government) as the initiative that would redeem the Northeast, given the economic stagnation of the region. The aim, in the words of the federal government of the time, was to correct the region's marked economic imbalance, offering the basic conditions for unleashing the process of industrialisation in the Northeast (SÁ; BRASIL, 2005).

The construction of the first Paulo Afonso hydroelectric power station (PA - I) was highly questioned, with regionalist speeches both for and against the implementation of hydroelectric projects in Paulo Afonso. Not only were there speeches of dissatisfaction from the south and southeast, against the construction of the plants, but also from the northeast, against the very board of Chesf, which was made up only of engineers from the "south of the country" as it was called at the time (SANTOS, 2008).

According to Souza (1955), these speeches began in the capital of Pernambuco and were intended to discredit the company's management and services. According to the author, these dissatisfactions presented Chesf's board of directors as a group of evildoers, organised to harm the Northeast, since these directors were seen as enemies of the Northeast and plunderers of the region's people.

As such, Castro (1992, p. 41) refers to the political nature of regional interactions, stating that this process:

> [...] they presuppose internal identification and cohesion and external competition for the defence of standards, preservation or obtaining more advantageous conditions. Thus, the regionalist character is both intrinsic and relative, [represented by] some level of regional tension, latent or manifest.

However, the intrinsic scale of the regional tensions referred to by Castro (1992) can be seen in the processes that led to the dismemberment of the then district of Paulo Afonso in 1959 from the neighbouring municipality of Gloria/BA. These tensions came above all from representatives of local political parties who protested against the construction of the wall erected by Chesf to separate the camp, which the company had built, from "Vila Poty", which was the place where the unskilled labourers lived and who felt, in a way, discriminated against in terms of access to the services and infrastructure that the camp offered (FOLHA SERTANEJA, 2015).

After the pioneering phase of building the first Paulo Afonso PA I Hydroelectric Power Station, the expansion and consolidation phase of Chesf's hydroelectric system followed, with several power stations being built, while the power transmission system was rapidly expanded, mainly to serve the northeastern capitals (SANTOS, 2010).

Therefore, the original project to build PA I foresaw the possibility of expanding the Paulo Afonso hydroelectric complex, so in 1955 work began on the Paulo Afonso II hydroelectric plant, which had its last generators installed in 1968. The increase in demand, however, led to the construction of the Paulo Afonso III hydroelectric plant, which began in 1966 and was completed and fully installed in 1974. Like PA I, PA II and PA III were built in underground galleries. In 1974 they totalled 1,524 MW of installed capacity (CHESF, 2017).

For Santos (2008), the move of the headquarters of the São Francisco Hydroelectric Company (CHESF) from Rio de Janeiro to Recife in 1975 began the restructuring of the regional hydroelectric system. The installation of electro-intensive industries in the north-east led to an exponential increase in the demand for electricity, mainly due to the rise in the price of oil, the economic effects of which were felt above all from the second half of the 1970s onwards.

Consequently, it was in this context that work began on the Moxotó hydroelectric power station in 1971. Despite being located in the neighbouring municipality of Glória/BA, this hydroelectric plant is part of the Paulo Afonso hydroelectric complex. It was completed in 1974 with the filling of a reservoir with a capacity to accumulate 1 million cubic metres of water (CENTRO DE MEMÓRIA DA ELETRICIDADE NO BRASIL, 1998), flooding the urban area and part of the rural area of the municipality of Glória/BA (MENDONÇA; BRITO, 2007).

According to ANEEL (2005), from a technical point of view, the Moxotó hydroelectric plant not only aimed to increase the strong demand for energy in the northeast, but also to regularise the flow of the São Francisco River in the Paulo Afonso region on a multi-weekly basis.

However, from a socio-economic and environmental point of view, the filling of the reservoir at the Apolônio Sales power station caused great harm to the local population. The resettlement of the population was rushed, with both urban and rural residents forced to leave the area quickly so that the reservoir could be filled (NOBRIGA, 2011; IULIANELLI, 2000).

On the other hand, at the time, a new headquarters was built for the municipality of Glória/BA, in an environment in which the population had little say over its location, and part of the residents of the urban area of that town ended up resisting staying in the newly built urban centre, preferring to set up an alternative centre located on the edge of the hydroelectric dam lake, in the place called Quixabá, which is approximately 30 kilometres upstream from the Paulo Afonso/BA hydroelectric complex (REIS, 2004).

The changes to the landscape were significant. The city was rebuilt with standardised housing, similar to large housing estates. The change in the urban landscape in this case tended to change people's social representations of the new city. It can also be said that cultural values were substantially affected, changing the meanings and affective value that resettled people placed on their spaces, their traditions, habits and customs (SANTOS, 2007).

The new seat of the municipality of Glória, due to its proximity to the urban area of the municipality of Paulo Afonso/BA and its regional importance (service hub, base of the São Francisco hydroelectric complex and aggregator of skilled labour), especially in the service sector (public services, retail and wholesale trade), ended up losing the typical functions inherent in the urban space of a city. In this sense, the city's urban functions ended up being transferred to the more important regional centre (CODEVASF, 2001; BATISTA, 1999).

As a result, the municipality's socio-economic base was disorganised without any conditions being provided for its reorganisation, mainly affecting farming activities, with the flooding of the fertile alluvial flats along the riverbank and the loss of space for the breeding of some herds, which were raised close to the fields along the riverbank (MENEZES; ROCHA, 2010). Many of the indemnities were made at a very low cost, since the vacant lands were disregarded, i.e. those that were not titled, in which case only those with title deeds were indemnified. This fact, according to Andrade (1993), led to court challenges over the amounts compensated, and in some cases these amounts were corrected by more than 100 per cent.

Parallel to the construction of the Apolônio Sales hydroelectric plant, construction began on the Paulo Afonso IV hydroelectric plant in 1972, the last of the hydroelectric plants that make up the Paulo Afonso complex. Its construction was based on the possibility of harnessing the waters of the Apolônio Sales reservoir by building a canal more than 5 kilometres long and with an average width of 135 metres. After filling the small reservoir that formed around the city of Paulo Afonso/BA, the process of urban occupation of the spaces left over from the formation of that lake intensified, resulting in the strong expansion of the city's peripheral neighbourhoods, as well as the use of its banks for leisure, with the formation of river beaches and small properties, "leisure farms" that lend themselves to the leisure of the local population. This hydroelectric plant had its power generation capacity fully completed in 1983, when it accumulated a total installed capacity of 2,460 MW (OLIVEIRA, 2005).

Finally, even before the Paulo Afonso IV hydroelectric plant was completed, the need to regularise the São Francisco River's water regime was already known, given the great variation observed between periods when the volume was very low and those when there was a surplus of water. This problem was overcome with the construction of the Sobradinho Hydroelectric Power Station, located 40 kilometres upstream from the municipalities of Petrolina/PE and Juazeiro in Bahia. Its construction was completed in 1979 with a final installed capacity of 1059 MW. Also noteworthy were the socio-environmental impacts produced by the construction of the dam, which required the relocation of around 64,000 people (PEREIRA, 2012; MUCCINI; MALTA, 2007).

CHAPTER 7

FINAL CONSIDERATIONS

The research highlighted the process that triggered the urban-regional evolution of the municipality of Paulo Afonso/BA, where the hydroelectric dams were built on the lower reaches of the São Francisco River. In this context, it was based on the evidence that the development project for modern Brazil, which began in the 1940s, required the construction of macro-structures to support the development of the northeastern industrial park through the generation and distribution of electricity.

Furthermore, the area where the municipality of Paulo Afonso was formed at the end of the 1940s was chosen for its great potential to generate hydroelectric power, although the distance from the major centres was an obstacle that was ultimately overcome by the geographical originality and physiographic difference that the region represented.

In turn, the exploitation of the hydroelectric potential of the São Francisco River in the semi-arid stretch of northeastern Brazil represented for the northeastern society of the first decades of the 20th century who lived in the area, a process of introducing a model of inducing development that was widespread at the time, that is, development reduced to the idea of simple economic growth.

Today, the municipality of Paulo Afonso/BA has a planned infrastructure that has been designed since its creation. The centre is located on an artificial island that was built with the implementation of the Paulo Afonso IV power plant canal (P.A.IV), around the centre are some neighbourhoods, outside the island there are also other important neighbourhoods, such as the BTN (Bairro Tancredo Neves I, II and III) which is the most populous neighbourhood, further away are the rural areas and districts of the city.

According to IBGE data (2017), Paulo Afonso/BA has one of the highest GDPs (Gross Domestic Product) in the state of Bahia and in 2007 it was on the list of the 30 largest cities in the northeastern interior with the highest GDPs. Today, it is among the 20 largest, occupying seventh place, with a GDP of 2,037,815 million reais.

Finally, it is worth emphasising that the municipality is also renowned for its natural beauty, such as the waterfalls and the Raso da Catarina environmental reserve (where endangered species such as the Lear's macaw and the eagle dove can be found), as well as for being home to the Chesf power station complex.

REFERENCES

ACSELRAD, H. Surveillance and Unity: the Urban Sustainability Agenda? VeraCidade Magazine. Year 2, n° 2, 1-11. 2007.

ALBERTI, V. Manual of oral history. Rio de Janeiro: FGV, 2008.

ANDRADE, M. C. de. Tradition and Change. The organisation of rural and urban space in the sub-medium São Francisco irrigation area. Rio de Janeiro: Zahar, 1993. p. 27-28.

ANDION, C. et al. The debate on the plural economy and its contribution to the study of sustainable territorial development dynamics. In: Eisforia. ano 4. volume 4. n. especial. Florianópolis: UFSC. 2007.

ANEEL, National Electric Energy Agency. BIG - Generation Information Bank. 2012. Available at: http://www.aneel.gov.br/aplicacoes/capacidadebrasil/capacidadebrasil.asp. Accessed on: 10 Apr. 2012.

National Electric Energy Agency. Atlas of electricity in Brazil. 2. Ed. - Brasília: ANEEL 2005.

BARBOSA, J. R., Class Notes - Flow Machines. Technological Institute of Aeronautics, São José dos Campos, Brazil, 2010.

BARDIN, L. Content analysis. São Paulo: Edições 70, 229p. 2011.

BATISTA, E. Nós Fizemos Paulo Afonso. Paulo Afonso, 1999.

BAPTISTA, S. G.; CUNHA, M. B. da. User research: an overview of data collection methods. Perspectivas em Ciência da Informação, v. 12, n. 2, p. 168-184, May/Aug. 2007. Available at: http://www.scielo.br/pdf/pci/v12n2/v12n2a11.pdf. Accessed on: 01 October 2017.

BELLET SANFELIU, C.; LLOP TORNÉ, J. M. The production of public space in Presidente Prudente: reflections from the perspective of gated subdivisions. In SPÓSITO, E.S., SPÓSITO BELTRÃO, M.E., SOBARZO, O. (eds). Cidades médias: produção do espaço urbano e regional, (199-214). São Paulo: Expressão popular. 2006.

________________. Miradas a otrosespacios urbanos: lasciudades intermedias. Scripta Nova,

Electronic Journal of Geography and Social Sciences. University of Barcelona, Spain, 165. Available at: (Consulted on: 10/10/2017). 2004.

BONDUKI, N.; ROLNIK, R. Periferia da Grande São Paulo: reprodução do espaço como expediente de reprodução da força de trabalho. In MARICATO, E. (comps.) A produção capitalista da casa (e da cidade) do Brasil industrial. São Paulo: Alfa-ômega. 1982.

BRANDÃO, Z. Between questionnaires and interviews. In: NOGUEIRA, Maria Alice; ROMANELLI, Geraldo; ZAGO, Nadir (Org.). Família e escola: trajectórias de escolarização em camadas médias e populares. Petrópolis: Vozes, p. 171- 183. 2003.

BRAZIL. Ministry of the Environment. Water Law - Law 9433, of 8 January 1997. Water Resources: set of legal norms. 3. ed. Brasília: Ministry of the Environment/Secretariat of Water Resources, 2004.

.CONAMA RESOLUTION No. 237, of 19 December 1997.

.CONAMA RESOLUTION NO. 006. Published in the Official Gazette of 22 October 1987, Section I, Page 17,499. 16 September 1987.

BRITO. M. C. W. de. Conservation Units: intentions and results. Ed. Annablume: FAPESP, São Paulo - SP, 230 p. 2001.

CALDEIRA, T. P.R. Cidade de muros: crime, segregation and citizenship in São Paulo. São Paulo: Edusp. 2000.

CANO, W. Regional Productive Deconcentration in Brazil 1970-2005. São Paulo: Editora UNESP. 2008.

CARLOS, A. F. A. Space as a condition for reproduction. The spatial condition. São Paulo: Contexto. 2011.

CARLOS, A. F. A., SOUZA, M. L., SPÓSITO, M. E. B. (Orgs.). The production of urban space: agents and processes, scales and challenges, (123-145). São Paulo: Contexto. 2013.

CARVALHO, E. Exclusão social e crescimento das cidades médias brasileiras. Scripta Nova - Revista electrónica de geografía y cienciasociales. University of Barcelona, VII 146. 2003.

CASSELL, C.; SYMON, G. Qualitative methods in organisational research. London: Sage Publications, 1994.

CASTELLS, M. O poder da identidade. São Paulo: Paz e Terra, 2008.

___________. The Urban Question. Rio de Janeiro: Paz e Terra. 1983.

CASTRO, I. E. de. The Myth of Necessity: Discourse and Practice of Northeastern Regionalism. Rio de Janeiro: Bertrand Brasil, 1992.

CHESF. PAULO AFONSO HYDROELECTRIC COMPLEX,   BAHIA. https://www.chesf.gov.br/SistemaChesf/Pages/SistemaGeracao/ComplexoPauloAfonso.aspx. Accessed on 05 November 2017.

______. São Francisco Hydroelectric Company. Recife, 2002.

CME; ELECTRICITY MEMORY CENTRE. 50 ANOS CHESF - 1948/1998, Rio de Janeiro: 1998.

______. Memória da eletricidade - Rio de Janeiro: FGV, 1993.

CODEVASF. Almanac - São Francisco Valley. 1st ed. Brasília, 2001.

COELHO, M. C. N.; CUNHA, L. H. Environmental Policy and Management. In: Cunha, Sandra B.; GUERRA, Antonio J.T. A questão ambiental: diferentes abordagens. Rio de Janeiro: Bertrand Brasil, 2003, p 43-79.

CONSTANZA, R. et al. The value of the world's ecosystem seviches and natural capital. Nature. (387). p. 253-260. 1997.

CORRÊA, R. L. Residential segregation: social classes and urban space. In: VASCONCELOS, P. A.; CÔRREA, R. L.; PINTAUDI, S. M. (Comps.). The contemporary city: spatial segregation (39-59). São Paulo: Contexto. 2013

________________. Urban Space. São Paulo: Ática. 1989.

COSTA, A. Metodologia científica. Mafra: Nosde, 2006.

ERIKSON, E H. Identity: youth and crisis. Rio de Janeiro: Zahar, 1987.

FACHIN, O. Fundamentos de metodologia. 4. ed. São Paulo: Saraiva, 2003.

FAHRIG, L. How much habitat is enough? Biological Conservation. vol. 100. p. 65- 74. 2001.

FERRAZ, O. M. Paulo Afonso Hydroelectric Power Station. In_ Águas e Energia Elétrica Nº 2, October 1999.

FLICK, U. Introduction to qualitative research. Porto Alegre/RS. Artmed. 2009. 405p.

______. Quantitative methods in scientific research. Translation by Artur M. Parreira. Lisbon: Monitor, 2005.

SERTANEJA LEAF. The energy of Paulo Afonso has been changing the history of the Northeast for 60 years. News_21866407. Paulo Afonso/BA. Folha Sertaneja online. 9p.   2015. Available at: http://www.folhasertaneja.com.br/noticia/21866407/especiais/a-energia-de-paulo-afonso-muda-a-   historia-do-nordeste-ha-60-anos/?indice=10. Accessed on: 19/01/2018.

FONSECA, R. C. How to write research projects and monographs – Practical Guide. Curitiba: Official Press, 2007.

FURNAS. 1957-1967: how it all began. Furnas Magazine. Special Edition 50 anos de Furnas. n.337, Rio de Janeiro, Feb./ 2007.

GAMEIRO, L. SHPPs lose ground to wind turbines in the energy market. 2012.

GARCIA, R. Interdisciplinarity and Complex Systems. In: LEFF, E. Ciencias Sociales y Formación Ambiental. Barcelona. Gedisa. 1994.

GIL, A. C. Métodos e técnicas de pesquisa social. 6 ed. São Paulo: Atlas, 2009.

GOMES, F. de A. M. História & energia: a eletrificação no Brasil. São Paulo: Eletropaulo, 1986.

GÓMEZ S., CÉSAR J. El fragmento urbano residencial enlaconstrucción de lametrópoli barcelonesa (1976-2006). Doctoral Thesis. Escuela Técnica Superior de Arquitectura de Barcelona - Universidad Politécnica de Cataluna, Barcelona - Spain. 2011.

HABERMAS, J. The Logic of the Social Sciences. São Paulo, Vozes, 2009.

HAESBAERT, R. O mito da desterritorialização. Do fim dos territórios à multiterritorialidade. 2. ed. Rio de Janeiro: Bertrand Brasil, 2006.

. De-territorialisation and identity: *the* "gaucho" network in the northeast. Niterói: EdUFF, 1997.

HALFELD, G. F. Atlas e Relatório Concernente a Exploração do Rio São Francisco. From Cachoeira da Pirapora to the Atlantic Ocean. Drawn up by order of the Government of H.M.I. Dom Pedro II in 1852, 1853 and 1854. Chesf Archive. Rio de Janeiro:

Litographia imperial.1860. Contains Iconography of the Author.

HALL, S. Cultural identity in postmodernity. Rio de Janeiro: DP&A, 2006

HAMBUS, I. Small Hydroelectric Power Plant.   2012. Available at: http://www.cimentoitambe.com.br/massa-cinzenta/pequena-central-hidreletrica- %E2%80%93- pch/.

HARVEY, D. Justice, nature, and geography of difference. Oxford: Blackwell Publishers, 1996.

__________. The Post-Modern Condition. Editora Loyola, São Paulo, 1993.

HOGAN, D. J. The relationship between population and the environment: challenges for demography. In: COSTA, H., TORRES, H. (Org.) Population, environment: debates and challenges. São Paulo: SENAC, p.21-52. 2000.

IBGE, Directorate of Research, Coordination of Population and Social and Physical Indicators. Estimates of the resident population with reference date 1 July 2017.

Research Directorate, Population and Social and Physical Indicators Coordination. Estimates of the resident population with reference date 2010. 2010.

IULIANELLI, J. A. S. A (short) analysis of the (recent) confrontations in the trade union centre of the São Francisco Sub-Medium : when the enemy is diffuse and criminal. Cadernos do CEAS, Salvador, n. 185, p. 37-56, Jan.-Feb. 2000.

JOLLIVET, M.; PAVE, A. "The environment: issues and perspectives for research". IN: VIEIRA, P.F. and WEBER, J.(organisers). Renewable resource management and development: new challenges for environmental research. São Paulo: Cortez. 1997.

JORDÃO, S. M. S. Management of lianas in edges of semideciduous seasonal forest and cerradão, Santa Rita do Passa Quatro, SP. Piracicaba, 2009. 248 f., il. Thesis (Doctorate in Sciences) - Luiz de Queiroz College of Agriculture, Piracicaba, 2009.

JUCÁ, J. 35 YEARS OF CHESF HISTORY . Recife: Chesf/Fundaj 1982.

KOWARICK, L. A. Urban Spoliation. Rio de Janeiro, Paz e Terra. 1979.

JUNIOR GRAMULIA, J., Contribution of the Henry Borden Hydroelectric Power Plant to the Operation Planning of Hydrothermal Power Systems. M.Sc.

Thesis, Universidade Federal do ABC, Santo André, SP, Brazil, 2009. 178f.

LEMOS JÚNIOR, C. B. THE IMPLEMENTATION OF THE FURNAS HYDROELECTRIC PLANT (MG) AND ITS REPERCUSSIONS: A STUDY ON THE TERRITORIALISATION OF PUBLIC POLICIES. 2010. 129f. Dissertation (Master's in Geography) - Institute of Geoscience, State University of Campinas, Campinas, 2010.

LOUREIRO, C. F. B. Sustainability and education: a look at political ecology. São Paulo: Cortez, 2012.

______________. The environmental movement and critical thinking: a political approach. Rio de Janeiro: Quartet, 2003.

LUCO, C. A.; VIGNOLI, J.R. Segregación residencial en áreas metropolitanas de América Latina: magnitud, características, evolución e implicaciones de política, 47. Santiago del Chile: CEPAL. 2003.

MARCONI, M. A.; LAKATOS, E. M. Metodologia científica. 5 ed. São Paulo: Atlas, 2009.

MARICATO, E. O Ministério das Cidades e a política nacional de desenvolvimento urbano. Políticas sociais - acompanhamento e análise, n.12, (211-220) Brasília: IPEA. 2006.

MARCOMIM, F. E; SILVA, A. D. V. The sustainable lightness of the university. In: GUERRA, A. F. S.; FIGUEIREDO, M. L. (Org.). Sustainabilities in dialogues. Itajaí: Univali, 2010. p. 171-189.

MARQUES, E. Social structure and segregation in São Paulo: transformations in the 2000s. In Revista Ciências Sociais, 57(3), 675-710. 2014.

MEDEIROS, V. A. S. UrbisBrasiliae ou sobre cidades do Brasil: inserindo assentamentos urbanos do país em investigações configuracionais comparativas. Thesis (Doctorate) Faculty of Architecture and Urbanism, University of Brasília: Brasília. 2006.

MEMORIAL UNIVERSITY OF NEWFOUNLAND (MUN). Canada Environmental Impact Assessment. EA Concepts. Available at: < http://www. ucs.mun.ca>. Accessed on: 19 August 2004.

MENDONÇA, L. L. de; BRITO, M. E.. (coord.) Paths of modernisation: chronology of electricity in Brazil (1878-2007). Rio de Janeiro: Centre for the Memory of Electricity in Brazil (1979-2007), 2007.

MELLO, F. Pernambucano de. Delmiro Gouveia: development driven by environmental preservation.

Recife: CHESF, Fundaj, Ed. Massangana, 1993.

MENEZES, A. C. S.; ROCHA, F. (Orgs.). A Resistência à Transposição do Rio São Francisco na Paraíba Historias de Luta em Defesa da Terra, das Águas e dos Povos do Nordeste. João Pessoa: Sal da Terra, 2010. 76 p.

MELAZZO, E. Mercado Imobiliário, expansão territorial e transformações intraurbanas: o caso de Presidente Prudente-SP/1975-1990. Rio de Janeiro: Federal University of Rio de Janeiro, Dissertation (Master's in Economics). 1993.

MIELNIK O.; NEVES C. C. Characteristics of the structure of hydroelectric power production in Brazil. In: ROSA, L. Pinguelli et al. Impacts of Large Hydroelectric and Nuclear Projects. Economic, Technological, Social and Environmental Aspects. São Paulo: Marco Zero, 1988. p. 1738.

MILARE, E. Prior environmental impact study in Brazil. In: AB"SABER, Aziz Nacib; PLANTENBERG, Clarita Muller (Orgs.). Impact forecasting. 2ª edition. São Paulo: Edusp, 2006. p. 51-83.

MILLENNIUM ECOSYSTEM ASSESSMENT. Ecosystems and Human Well-Being: Synthesis. Island Press, Washington, DC, 2005.

MINAYO, M. C. S. O desafio do conhecimento científico: pesquisa qualitativa em saúde. 2. ed. São Paulo: Hucitec-Abrasco, 1994.

MORET, A. S.; FERREIRA, I. A. Madeira HPPs: planning to meet electricity demand, speed of studies and socio-environmental consequences. In: BRAZILIAN ENERGY CONGRESS, 12, 2008, Rio de Janeiro. Proceedings... Rio de Janeiro: COPPE, UFRJ, 2008.

MORSELLO, C. Public and Private Protected Areas: selection and management. Ed. Annablume: FAPESP, São Paulo - SP, 344 p. 2001.

MOTA, C. N. da. Before (and after) the São Francisco River: the true discoverers of the Opara. In: SÁ, Antônio Fernando de Araújo; BRASIL, Vanessa Maria (Orgs.). River Without History? Readings on the São Francisco River. Aracaju: Fapese, 2005. p. 91-104.

MUCCINI, S.; MALTA, S. PIONEIRO PERÍODO DA HIDRELÉTRICA DE PAULO AFONSO- BA: UMA CONTRIBUIÇÃO À HISTORIOGRAFIA DE BASE LOCAL E REGIONAL. Rios Eletrónica. Scientific Journal of FASETE. Year 1. N° 01. August 2007.

MULLER, A. C. Hidrelétricas, meio ambiente e desenvolvimento. São Paulo: Makron Books, 1995.

NASCIMENTO, L. F. M. Paulo Afonso: light and force moving the Northeast. Salvador: EGBA/ACHÉ, 1998.

NETTO, V. M. What is space syntax not? Arquitextos, São Paulo, year 14, n. 161.04, Vitruvius, . Consultation: 10/04/2017). 2013.

NOBREGA, R. S. The Dam-Affected: refugees from an unknown war. Interdisciplinary Journal of Human Mobility. Brasília, n. 36, p. 125-143, jan./jun. 2011.

OLIVEIRA, R. R.. CHESF and the role of the state in electricity generation. 6 Revista Econômica do Nordeste , Fortaleza, v. 32, n. 1, p. 10-35, jan.-mar. 2001.

OLIVEIRA, E. A. F. de. On the Tracks of Piranhas History: an essay on the Paulo Afonso Railway. In: SÁ, Antônio Fernando de Araújo; BRASIL, Vanessa Maria (Orgs.). River Without History? Readings on the São Francisco River. Aracaju: Fapese, 2005. p. 221-239.

PEREIRA, M. da C. M.. "Local and Regional History - singularities of a plural history". In: FARIAS, Sara Oliveira; LEAL, Maria das Graças de Andrade (Orgs). Regional and Local History II: the plural and the singular in debate. Salvador, EDUNEB, 2012.

PEREIRA, R. H.M., BARROS, GONÇALVES A.P., HOLANDA, F.R.B., MEDEIROS, V.A.S. The use of Space Syntax in urban transport performance: limits and potential. Text for Discussion 1630. IPEA: Brasília. 2011.

; et al. Evaluation of the socioeconomic impacts of energy projects – rural electrification: ENERSUL concession area - MS. University of Vale do Rio dos Sinos. São Leopoldo: UNISINOS, 2005.

PESAVENTO, S. J. Visible cities, sensitive cities, imaginary cities. Revista Brasileira de História, vol. 27, n° 53, June 2007.

POMPEU, C. T. Water Law in Brazil. São Paulo: Ed. RT, 2006.

RAFFESTIN, C. For a Geography of Power. France. São Paulo: Ática, 1993.

RAMPAZO, A. V.; ICHIKAWA, E. Y. Shipwrecked identities: the impact of organisations on the (re)construction of the symbolic universe of the Salto Santiago riverside dwellers. Cad. EBAPE.BR, v. 11, n° 1, Mar. p. 104-127. 2013.

RAUEN, F. J. Roteiros de investigação científica. Tubarão: Unisul, 2002.

REIS, R. R. do A. PAULO AFONSO E O SERTÃO BAIANO: SUA GEOGRAFIA E SEU POVO. 1 ed. Paulo Afonso-BA: Fonte Viva, 2004.

RIBEIRO, L. C. Q. Dos cortiços aos condomínios fechados: as formas de produção da moradia na cidade do Rio de Janeiro. Rio de Janeiro: Brazilian Civilisation. 1997.

RODRIGUES, C.; ADAMI, S. Hydrography techniques. In: VENTURI, Luiz A. Bittar (Org.) GEOGRAPHY: Field, Laboratory and Classroom Practices. São Paulo: Editora Sarandi, 2011. p. 55-82.

ROLNIK, R. The logic of disorder. In: Lê Monde Diplomatique Brasil. Year 2 n° 13, August 2008, pp. 10-19. São Paulo. 2008.

ROMERA E SILVA, P. A. Água: quem vive sem? 2.ed. São Paulo: FCTH/CT-Hidro (ANA, CNPq/SNRH), 2004.

ROSA, L. P. Hydroelectric, thermoelectric and nuclear generation. Estudos Avançados. v. 21, n.59, p.39- 58. 2007.

ROSA, L. P., et al. A prospective evaluation of the National Electricity Conservation Programme - PROCEL. Report 1: Analysing electricity conservation in Brazil and PROCEL's activities. Rio de Janeiro: COPPE; PPE, 65. 1994.

ROSENTAL, Cl.; FRÉMONTIER-MURPHY, C. Introduction to quantitative methods in the humanities and social sciences. Porto Alegre: Instituto Piaget, 2001.

ROSS, J. L. S. Hydroelectric plants and their socio-environmental impacts. In: STIPP, Nilza A. F. (Org). Análise ambiental - usinas hidrelétricas: uma visão multidisciplinar, Londrina:UEL: NEMA, 1999. p.17- 27.

RUWER, L. M. E. Proposal for Self-Sustainable Planning Requirements for the Absorption of Graduates from Large Government Projects. Master's dissertation, Postgraduate Programme in Production Engineering, Federal University of Santa Catarina. 2004.

SÁ, A. F. de A.; BRASIL, V. M. (Orgs.). River Without History? Readings on the São Francisco River. Aracaju: Fapese, 2005.

SANCHES, L. A. U. The geography of energy in Brazil - part 1: from the Empire to the

Estado Novo. Geografia, São Paulo, Issue no. 38, p. 38-47, July 2011.

SALLES, C. M. Rios e Canais. Editora Elbert, Florianópolis. 1993.

SANT'ANA, M. M. de. Annotated bibliography of Delmiro Gouveia, 1917-1994. Chesf, Recife: 1996, p. 19.

SANTIN, J. R.; FLORES, D. H. The historical evolution of the municipality in Brazilian federalism, local power and the city statute. Revista Justiça do Direito, Passo Fundo, v. 20 n°, 2006.

SANTOS, R. G. dos. Gouveia dos. Socio-environmental Impacts on the Banks of the São Francisco River: The Relationship between Man and Nature. São Paulo: Editora Biblioteca 24 horas, 2010. 192 p.

______________. Socio-environmental impacts on the banks of the São Francisco River: a case study. 2008. 193 f. Master's dissertation - Dpgeo/Fflch-USP, São Paulo, 2008.

SANTOS, S. Lament and pain: A socio-anthropological analysis of compulsory displacement caused by the construction of dams. 2007. 279f. Dissertation (Doctorate in Social Sciences) - Institute of Philosophy and Human Sciences, Federal University of Pará, Belém, 2007.

SANTOS. A. M. Medium-sized cities: new frontiers of opportunity. In ANGELA M.S., PENALVA S., GLAUCIO J.M., SANT'ANNA. M.J.G. (Org.). Rio de Janeiro: Um território em mutação. (47-72). Rio de Janeiro: Gramma. 2011.

______________. Urban Policy: the importance of focussing on medium-sized cities. Revista de Direito da Cidade, Rio de Janeiro, v.05, n. 02, 153- 177. 2010.

SANTOS, V. L. dos. Social impacts of large hydroelectric projects. Geography Notebook. Belo Horizonte, v.12, n.19, p.35-48, 2nd Semester, 2002.

SCARPIN, P. Once upon a time there was transposition. Piauí 75, [s. n.], issue no. 75, year 7, p. 28-34, December 2012.

SCHAEFFER, R. Environmental Impacts of Large Hydroelectric Plants in Brazil. Master's dissertation, Nuclear Engineering and Energy Planning, COPPE/UFRJ. 1986.

SEVÁ, O. Strange cathedrals: notes on hydroelectric capital, nature and society. Ciência e Cultura, São Paulo, v. 60, n.3, p.44-50, September 2008.

SEVÁ FILHO, A. O. Critical knowledge of mega-hydroelectric dams: how else to evaluate natural alterations, social transformations and the destruction of river monuments. In: ANPPAS NATIONAL MEETING, 2, 2004. Electronic Proceedings... Indaiatuba, SP. Available at: www.anppas.org.br.

SILVA, R. M. da. Between Paris, Rio de Janeiro and Bahia: Itaberaba's internal policy to modernise and civilise itself. Monograph (Degree in History). Itaberaba: UNEB, 2013.

__________. Urban restructuring and socio-spatial segregation in the interior of São Paulo In Scripta Nova, Revista electrónica de geografía y cienciassociales, XI 245(11). University of Barcelona, Barcelona, 2007.

SOBARZO, O. Socio-spatial segregation in Presidente Prudente: an analysis of horizontal condominiums. Master's dissertation. Presidente Prudente: UNESP. 1999.

SPÓSITO, M. E. B. The production of urban space: scales, differences and socio-spatial inequalities. In CARLOS, A. F. A., SOUZA, M. L., SPÓSITO, M. E. B. (Orgs.). The production of urban space: agents and processes, scales and challenges, (123-145). São Paulo: Contexto. 2013.

SPÓSITO, M. E. B. Capitalism and Urbanisation. 16ª ed. São Paulo: Contexto. 2012.

SOUZA, E. B. C A Geopolítica da produção do Espaço: localização da hidrelétrica da Itaipu Binacional. Revista Geografares, n.9, p.141-167, Jul./Dec., 2011.

SOUZA, É. J. C. de. Territorial policies in the state of Bahia: regionalisation and planning. Dissertation (Master's in Geography). Salvador: UFBA, 2008.

SOUZA, A. J. A. de. The Energy of Paulo Afonso and the Northeast. Recife: Gráfica e Editora do Recife S/A, 1955. 42 p.

TEIXEIRA, C. History of Energy in Bahia. Salvador: EPP Publicações e Publicidade, 2005.

VASCONCELOS, P. A. Contribution to the debate on socio-spatial processes and forms in cities. In VASCONCELOS, P. A.; CORRÊA, R. L.; PINTAUDI, S. M. (Orgs.). The Contemporary City: Spatial Segregation, (17-37). São Paulo: Contexto. 2013.

VIEIRA, P.F. and CAZELLA, A.A. Sustainable Territorial Development in rural areas: subsidies for the elaboration of an analysis model. In: Seminario Internacional Territorios Rurales en Movimiento, Santiago de Chile. Rural Territories on the Move: social movements, actors and institutions of rural territorial development. Santiago de Chile: IDRC-CRDI. 2006.

VIEIRA, A. B.; MELAZZO, E. Introdução ao conceito de segregação sócio espacial Revista Formação (Online), v. 1, nº 10, 2003. Available at: < http:// revista.fct. unesp.br/index.php/formacao/article/view/1118>. Accessed: 15/01/2016.

VILLAÇA, F. São Paulo: urban segregation and inequality. Estudos Avançados, v. 25, n. 71, 3758. 2011.

__________. Intra-urban space in Brazil. São Paulo: Studio Nobel, FAPESP, Lincoln Institute of Land Police. 2001.

ZHOURI, A.; OLIVEIRA, R. Development, social conflicts and violence in rural Brazil: the case of hydroelectric plants. Ambiente & Sociedade, Campinas, v.10, n.2, p.119-135, jul./dez. 2007.

yes
I want morebooks!

Buy your books fast and straightforward online - at one of world's fastest growing online book stores! Environmentally sound due to Print-on-Demand technologies.

Buy your books online at
www.morebooks.shop

Kaufen Sie Ihre Bücher schnell und unkompliziert online – auf einer der am schnellsten wachsenden Buchhandelsplattformen weltweit! Dank Print-On-Demand umwelt- und ressourcenschonend produzi ert.

Bücher schneller online kaufen
www.morebooks.shop